Assessment and Analysis of Waste Treatment and Environmental Management

Assessment and Analysis of Waste Treatment and Environmental Management

Guest Editors
Robert Oleniacz
Katarzyna Grzesik

Basel • Beijing • Wuhan • Barcelona • Belgrade • Novi Sad • Cluj • Manchester

Guest Editors

Robert Oleniacz
Department of Environmental
Management and Protection
AGH University of Krakow
Krakow
Poland

Katarzyna Grzesik
Department of Environmental
Management and Protection
AGH University of Krakow
Krakow
Poland

Editorial Office
MDPI AG
Grosspeteranlage 5
4052 Basel, Switzerland

This is a reprint of the Special Issue, published open access by the journal *Energies* (ISSN 1996-1073), freely accessible at: www.mdpi.com/journal/energies/special_issues/wy_em.

For citation purposes, cite each article independently as indicated on the article page online and using the guide below:

Lastname, A.A.; Lastname, B.B. Article Title. *Journal Name* **Year**, *Volume Number*, Page Range.

ISBN 978-3-7258-3120-3 (Hbk)
ISBN 978-3-7258-3119-7 (PDF)
https://doi.org/10.3390/books978-3-7258-3119-7

© 2025 by the authors. Articles in this book are Open Access and distributed under the Creative Commons Attribution (CC BY) license. The book as a whole is distributed by MDPI under the terms and conditions of the Creative Commons Attribution-NonCommercial-NoDerivs (CC BY-NC-ND) license (https://creativecommons.org/licenses/by-nc-nd/4.0/).

Contents

About the Editors vii

Preface ix

Robert Oleniacz and Katarzyna Grzesik
Assessment and Analysis of Waste Treatment and Environmental Management
Reprinted from: *Energies* **2025**, *18*, 138, https://doi.org/10.3390/en18010138 1

Boško Josimović, Božidar Manić and Nikola Krunić
Strategic Environmental Assessment as a Support in a Sustainable National Waste Management Program—European Experience in Serbia
Reprinted from: *Energies* **2022**, *15*, 4568, https://doi.org/10.3390/en15134568 11

Anna Gronba-Chyła, Agnieszka Generowicz, Paweł Kwaśnicki, Dawid Cycoń, Justyna Kwaśny and Katarzyna Graz et al.
Determining the Effectiveness of Street Cleaning with the Use of Decision Analysis and Research on the Reduction in Chloride in Waste
Reprinted from: *Energies* **2022**, *15*, 3538, https://doi.org/10.3390/en15103538 24

Sonja Cerar, Luka Serianz, Katja Koren, Joerg Prestor and Nina Mali
Synoptic Risk Assessment of Groundwater Contamination from Landfills
Reprinted from: *Energies* **2022**, *15*, 5150, https://doi.org/10.3390/en15145150 35

Piotr Jakubowicz, Teresa Steliga and Katarzyna Wojtowicz
Analysis of Temperature Influence on Precipitation of Secondary Sediments during Water Injection into an Absorptive Well
Reprinted from: *Energies* **2022**, *15*, 9130, https://doi.org/10.3390/en15239130 52

Jolanta Latosińska, Maria Żygadło and Marlena Debicka
The Biological Drying of Municipal Waste in an Industrial Reactor—A Case Study
Reprinted from: *Energies* **2022**, *15*, 1039, https://doi.org/10.3390/en15031039 69

Aleksandra Leśniańska, Beata Janowska and Robert Sidełko
Immobilization of Zn and Cu in Conditions of Reduced C/N Ratio during Sewage Sludge Composting Process
Reprinted from: *Energies* **2022**, *15*, 4507, https://doi.org/10.3390/en15124507 78

Arkadiusz Nedzarek, Małgorzata Bonisławska, Agnieszka Tórz, Adam Tański and Krzysztof Formicki
Effect of Filter Medium on Water Quality during Passive Biofilter Activation in a Recirculating Aquaculture System for *Oncorhynchus mykiss*
Reprinted from: *Energies* **2022**, *15*, 6890, https://doi.org/10.3390/en15196890 97

Shuping Pan, Qi Yao, Wenxiang Cai, Yaqi Peng, Yuhao Luo and Zhizhen Wang et al.
Characterization of Dioxins and Heavy Metals in Chelated Fly Ash
Reprinted from: *Energies* **2022**, *15*, 4868, https://doi.org/10.3390/en15134868 111

Monika Czop, Beata Łaźniewska-Piekarczyk and Małgorzata Kajda-Szcześniak
Evaluation of the Immobilization of Fly Ash from the Incineration of Municipal Waste in Cement Mortar Incorporating Nanomaterials—A Case Study
Reprinted from: *Energies* **2022**, *15*, 9050, https://doi.org/10.3390/en15239050 121

Kamil Banaszkiewicz, Tadeusz Marcinkowski and Iwona Pasiecznik
Fly Ash as an Ingredient in the Contaminated Soil Stabilization Process
Reprinted from: *Energies* **2022**, *15*, 565, https://doi.org/10.3390/en15020565 . . . . **137**

Beata Zygmunt-Kowalska, Kinga Pielichowska, Patrycja Trestka, Magdalena Ziabka and Monika Kuźnia
The Effect of Ash Silanization on the Selected Properties of Rigid Polyurethane Foam/Coal Fly Ash Composites
Reprinted from: *Energies* **2022**, *15*, 2014, https://doi.org/10.3390/en15062014 . . . . **152**

About the Editors

Robert Oleniacz

Dr. Robert Oleniacz is an assistant professor at AGH University of Krakow (formerly AGH University of Science and Technology in Krakow, Poland). He began working there in 1993, initially as a technician and later as an assistant in the Department of Environmental Management and Protection, part of the Faculty of Mining Surveying and Environmental Engineering (now titled Faculty of Geo-Data Science, Geodesy, and Environmental Engineering). He holds an MSc degree (1994) and a PhD degree (2001) in Environmental Engineering from AGH University of Krakow. Dr. Oleniacz has authored more than 180 research papers, including 80 articles published in peer-reviewed journals and over 100 chapters in scientific monographs or papers in conference proceedings. He has also delivered more than 40 presentations at scientific conferences. Dr. Oleniacz has supervised three PhD dissertations as an assistant supervisor, 115 master's theses, and nearly 230 engineering diploma projects. Moreover, he has completed over 90 research projects commissioned by industry or public administration. His most recent project, which he led, involved assessing the impact on air quality of a large-scale fire at a hazardous waste warehouse in western Poland. His scientific research focuses on waste management and air quality protection, encompassing areas such as evaluating the performance of selected waste management facilities (e.g., waste incineration plants, mechanical–biological treatment plants, and refuse-derived fuel production lines), measuring air pollutant emissions from hazardous waste incinerators, and assessing the air quality impact of waste incineration/co-incineration plants as well as uncontrolled or open waste incineration processes. Currently, a significant portion of his research activity involves identifying factors influencing air pollution in specific regions and enhancing or applying practical measurement and modeling methods for air quality impact assessment and management.

Katarzyna Grzesik

Katarzyna Grzesik is an associate professor at AGH University of Krakow, where she earned both her undergraduate degree in Environmental Engineering and her PhD. In 2018, she received her habilitation (postdoctoral degree). Her research interests focus on integrated municipal and industrial waste management systems, circular economy practices, and environmental life cycle assessment (LCA) of products and processes, with particular emphasis on applying LCA to waste management systems and assessing the environmental impact of advanced bio-waste fuel production. Additionally, she investigates waste fires and their environmental impacts. Professor Grzesik has published almost 100 research papers, including nearly 50 articles in peer-reviewed journals and 50 conference papers. She has also served as a reviewer for close to 50 articles in scientific journals within the JCR database. As an experienced independent expert for the European Commission, Professor Grzesik has participated in around 30 calls for Horizon 2020, Horizon Europe, and Era Net programs. She was the coordinator of the national grant "Waste Fires – Identification and Evaluation of Environmental and Climate Impacts", served as a Work Package leader in an international grant on rare earth element extraction from secondary sources, and contributed to various international grants. Her professional experience includes the development of numerous waste management plans for municipalities and the completion of many commissioned projects for companies and institutions, such as ArcelorMittal and SGL Carbon. Professor Grzesik has supervised one completed PhD dissertation and is currently supervising another.

Preface

A well-planned waste management system can offer numerous benefits to both the environment and the economy, with the advantages outweighing the incurred costs and remaining acceptable in terms of potential negative environmental impacts. Therefore, making optimal decisions about the system as a whole and its individual components is crucial. This requires the use of appropriate environmental management tools and research methods to assess detailed solutions. Many of these tools and methods have been developed and applied in the research papers presented in the Special Issue entitled "Assessment and Analysis of Waste Treatment and Environmental Management".

This Special Issue includes one editorial and eleven high-quality papers addressing diverse aspects of waste management and environmental protection. The topics covered range from the application of strategic environmental assessment to the development of national waste management programs to multi-criteria decision analysis for chloride removal in street-washing and cleaning processes. This Special Issue also examines the concept of synoptic risk assessment for landfills and their impact on groundwater, as well as the use of hydrogeochemical modeling to address challenges related to near-well zone colmatation in wastewater reuse after hydrocarbon extraction. Additional research focuses on the biological drying of municipal solid waste, the composting of sewage sludge, wastewater treatment and reuse in fish aquaculture, and the properties of fly ash from waste incineration and coal combustion, with a focus on innovative management methods. These studies contribute to the advancement of a circular economy and the zero-waste vision, aiming to eliminate waste and environmental pollution while preserving value in the form of energy, labor, and materials, thus conserving non-renewable raw materials. This Special Issue will be of interest to a broad audience, including but not limited to the following readers: researchers, students, plant operators, planners, and decision-makers involved in waste and sewage management.

Robert Oleniacz and Katarzyna Grzesik
Guest Editors

Editorial

Assessment and Analysis of Waste Treatment and Environmental Management

Robert Oleniacz * and Katarzyna Grzesik

Department of Environmental Management and Protection, Faculty of Geo-Data Science, Geodesy and Environmental Engineering, AGH University of Krakow, Mickiewicza 30 Av., 30-059 Krakow, Poland; grzesikk@agh.edu.pl
* Correspondence: oleniacz@agh.edu.pl

1. Introduction

The waste management sector plays a crucial role within the broader environmental management system. This sector impacts the environment both directly and indirectly, potentially causing adverse effects such as pollution of air, water, and soil. However, it also offers significant environmental benefits when waste recycling and energy recovery processes are conducted effectively. By implementing responsible waste management practices, this sector could contribute to reducing pollution and conserving resources through sustainable recycling and recovery initiatives [1–3]. Properly selected and managed waste processing methods should deliver economic benefits at reasonable environmental costs while also being socially acceptable. Achieving a balance between economic gains and ecological responsibility is essential for sustainable waste management practices [4,5]. The adoption of a circular economy enables the reuse of waste materials and the conservation of essential raw materials—including critical resources and non-renewable fuels—while also reducing the environmental impact across the entire product life cycle [6–8]. Focusing on recycling and waste recovery can help us to realize the vision of a zero-waste system, fostering changes within industrial systems, including manufacturing processes and the integration of Industry 4.0 technologies [8–10].

Certain market conditions may render the recycling or recovery of some waste unprofitable. This can lead to wasted materials and energy, the potential for serious environmental issues, and even the risk of failure or collapse within parts of the waste management system. Examples include directing waste fractions that can be recycled organically or materially to landfills or to incineration [11–14], improper storage or landfill of troublesome waste leading to soil and water environment pollution, uncontrolled biogas generation and greenhouse and odor gas emissions, or causing spontaneous combustion and fires in landfills [15–19]. This also includes the illegal disposal of flammable waste through deliberate burning by arson [20–23], as well as the open burning of agricultural residues or municipal solid waste [24–27] and the combustion of waste in residential furnaces [28–30].

The challenges encountered by waste management and environmental management systems differ across regions, influenced by factors such as economic development, residents' lifestyles and habits, and the effectiveness of local regulations, strategies, and legal enforcement [31–35]. Various approaches are available for decision-making in this area, ranging from basic to advanced methods. These include data-generation techniques (e.g., surveys), simple assessment methods (e.g., benchmarking), and more complex assessment methods (e.g., multi-criteria decision-making) [36]. However, it remains essential to conduct detailed studies on specific types of waste and sewage, as well as on the challenges associated with their processing, particularly regarding their environmental impact.

Received: 27 November 2024
Accepted: 26 December 2024
Published: 1 January 2025

Citation: Oleniacz, R.; Grzesik, K. Assessment and Analysis of Waste Treatment and Environmental Management. *Energies* **2025**, *18*, 138. https://doi.org/10.3390/en18010138

Copyright: © 2025 by the authors. Licensee MDPI, Basel, Switzerland. This article is an open access article distributed under the terms and conditions of the Creative Commons Attribution (CC BY) license (https://creativecommons.org/licenses/by/4.0/).

Depending on the waste treatment technology and the degree of the systemic approach, various methods can be used to assess waste management systems, including cost benefit, cost effectiveness, mass balance, and material flow analysis [37,38]; energy, exergy, or emergy analysis [37,39–41]; life cycle assessment [39,42–45]; and methods used directly in risk, social, environmental impact, or strategic environmental assessments [46–50].

An invitation to publish research results in a Special Issue (SI) titled "Assessment and Analysis of Waste Treatment and Environmental Management" was extended to numerous scientists working in this area and related fields. The substantial response led to the continuation of this topic in a subsequent SI [51–53]. A general review of the papers published in the first SI is provided in Section 2.

2. General Review of the Papers Published in the Special Issue

The Special Issue published 11 research papers that present assessments and analyses conducted by their authors in the field of waste management, focusing on various treatment processes and the subsequent use of waste. These studies are also significant in the context of environmental management and protection. While most of the research comprises case studies, some offer a more comprehensive approach to the issue, with a case study serving as an additional component. Below, the papers are grouped and broadly characterized, highlighting the most important research findings and conclusions.

In paper [54], an exemplary application of the Strategic Environmental Assessment (SEA) method was presented for the National Waste Management Program (NWMP) in the Republic of Serbia, a country that aligns with EU regulations on waste management and environmental protection. The program outlines strategic goals for improving the waste management system and establishes the fundamental principles that all participants in the system should follow to achieve these goals during the program's validity period (2022–2031). During the preparation stage for applying the SEA method, the authors concluded that this method relies on qualitative assessment and techniques, where expert assessment plays a crucial role, particularly, for assessing alternative scenarios. Based on the results of the performed multicriteria evaluation (a semi-quantitative method), appropriate actions were identified to mitigate potential negative effects during the implementation of the NWMP. The paper's conclusions emphasized that a clear matrix presentation is an effective way to present the results of multicriteria evaluation, especially in SEA phases that involve public participation. The findings can also serve as valuable data and guidelines for employing other assessment methods, including those at a lower level of impact assessment (e.g., Environmental Impact Assessment or Environmental Social Impact Assessment).

In paper [55], an assessment of chloride removal strategies in street washing and cleaning processes was conducted using multi-criteria decision analysis. The primary objective of the research was to determine the amount of chlorides used in street sweeping on an annual basis in a medium-sized city (approximately 55,000 inhabitants) and to identify potential environmental risks associated with their impact on selected environmental aspects. An additional goal was to assess the frequency of urban street cleaning, considering both environmental and economic effects. The area from which street cleaning waste is collected in the analyzed city spans approximately 150 km^2, with the annual amount of sweepings ranging from 300 to 1200 tons, depending on the severity of winter conditions. The study found that the highest chloride concentrations in street cleaning waste occurred during the winter months, with significant fluctuations based on the sampling location. Maximum concentrations of 1200–1700 mg/dm^3 were observed in waste from street and sidewalk cleaning around sewer gullies. Large quantities of these sweepings deposited in landfills can increase the salinity of landfill leachate, potentially delaying the decomposition of organic matter, promoting biogas formation, and increasing the risk of groundwater salinization. This can also hinder biological processes in sewage treatment plants. The multi-criteria analysis suggested that the most effective approach to street cleaning, and consequently the best method for reducing chloride concentrations in the waste, would involve both sweeping and daily washing.

Paper [56] presents a developed method for synoptic risk assessment based on a conceptual model of landfill and the results of groundwater chemical monitoring, which was tested on 69 landfills in Slovenia. The method proposed by the authors focuses on estimating the relationship between key chemical parameters, incorporating various conceptual models of landfills. The study recorded a wide range of field parameters, including inorganic elements, microelements, and organic compounds, as well as volatile organic compounds (VOCs), aromatic hydrocarbons (BTEX), triazine, and organochlorine pesticides in groundwater. The research revealed that most landfills in Slovenia can negatively impact groundwater quality, generally to a small or medium extent, though in some cases the impact is significant. For example, the classification system used in the study identified 24 landfills with a pronounced effect on groundwater quality, which may also influence the chemical status of the groundwater body; these were classified in Priority Class 1. Another 31 landfills were classified in Priority Class 2, while 14 were placed in Priority Class 3. The conclusions of the paper highlight that the developed classification system, based on synoptic risk assessment, is effective for evaluating the environmental impact of landfills and can serve as a guideline for similar assessments in other regions of the world.

The issues addressed in the next paper [57] focus on the management of wastewater produced during hydrocarbon extraction processes. These waters are highly contaminated with a wide range of chemical compounds that can have detrimental effects on the environment. The extraction and storage of water may also lead to the precipitation of secondary sediments, which can cause problems when injecting subsequent water batches with varying properties into absorbent layers. The paper describes innovative applications of a specialized hydrogeochemical modeling system (PHREEQC) to identify issues related to near-well zone clogging. It was found that changes in the solubility of various minerals, depending on the chemical composition and physical–chemical parameters of the water sample, can result in either precipitation or dissolution, with varying degrees of intensity. A key factor influencing these changes is temperature, particularly when the temperature of the water increases to the level of the reservoir. If the reservoir water is saturated with a mineral at a low temperature, heating may cause sediment precipitation. The formation of such sediments can reduce the permeability of reservoir rocks. The findings from this research can also be applied to environmental management practices, aiming to minimize contaminants released into the environment during oil and natural gas extraction.

Work [58] presents the results of municipal solid waste (MSW) treatment using the biological drying process in a full-scale reactor (150 m^3). Biodrying reduces moisture content and volume while enhancing bulk density through the effective utilization of biological heat. As a result, it can serve as one of the initial stages of waste processing in mechanical–biological treatment (MBT) plants for the production of refuse-derived fuel (RDF). During biodrying, organic matter undergoes partial decomposition by aerobic microorganisms. The paper compares and discusses the effects of 14 days of operation in the biodrying reactor, focusing on the daily variability in temperature both within the bed and in the air above it, as well as the daily fluctuations in waste moisture, loss on ignition (LOI), and net heating value (NHV) for MSW 0–80 mm fractions processed in the front, middle, and back sections of the reactor, as well as average daily values for the entire reactor. The study's results showed that 14 days of biodrying led to a reduction in moisture and LOI by up to 50% and 66%, respectively, and an increase in NHV by more than 45% (up to

9.6 MJ/kg) compared to initial values. These findings can help biodrying reactor operators optimize the drying process to produce RDFs with the highest NHV and lowest LOI. This can be achieved by controlling the reactor's ventilation rate and temperature within the bed, thus speeding up moisture removal from the MSW. In MBT plants, the final products are typically screened, allowing the removal of undersized fractions (e.g., 0–20 mm) that mainly consist of ballast, further improving the calorific value of the produced RDF.

Biodegradable waste fractions that are unsuitable for alternative fuel production can be subject to organic recycling through composting and used for soil fertilization or the reclamation of degraded areas. However, not all such wastes are equally suitable for this purpose due to the presence of toxic components. In paper [59], the results of research on the composting process of sewage sludge are presented, focusing on the transformation of chemical forms of two elements (Cu and Zn), which were found in high concentrations in the studied waste. Specifically, the changes in the content of these heavy metals in fractions separated through sequential extraction in compost samples were compared and analyzed, particularly for composts with reduced supplementation of barley straw (an additional source of organic carbon). The results indicated that the direction of transformation of the analyzed metals depended on the amount of barley straw added to the sewage sludge before the composting process. Reduced supplementation positively influenced the allocation of the tested elements in organic (IV) and residual (V) fractions, while decreasing the share of mobile heavy metal forms in the bioavailable fractions, particularly ion-exchangeable (I) and carbonate (II) forms. This study also used an artificial neural network (ANN) to develop a compost classification tool based on Austrian standards, which considers only the I ÷ IV fractions as the labile, potentially bioavailable portion of heavy metals bound in various chemical forms in compost. The developed ANN tool can predict the compost class (and quality) based on factors such as the C/N ratio, which is influenced by the proportion of the supplement providing additional organic carbon.

Other innovative studies focus on optimizing processes used for wastewater treatment and reuse in fish farming (aquaculture). Paper [60] presents the results of evaluating the passive activation of selected biofilter beds in a recirculating aquaculture system used for culturing cold-water salmonid fish (rainbow trout, *Oncorhynchus mykiss*). The study compares changes in the concentrations of nitrogen, phosphorus, and carbon during this process, considering the different shapes and specific surface areas of various plastic substrates. Commercial media with varying specific surface areas were used as substrates: a non-porous medium with a smooth surface (RK-Plast, 700 m^2/m^3), a hard porous medium (Mutag-BioChip30, 5500 m^2/m^3), and a soft porous medium with activated carbon (LevaPor, 2700 m^2/m^3). The experiments aimed to identify the most suitable biofilter bed for the water treatment system at the *O. mykiss* breeding facility under construction. The results showed that efficient nitrification occurred with all media variants, maintaining nitrate nitrogen concentrations and reducing the toxic forms of ammonium and nitrite nitrogen to levels safe for the cultured fish. The nitrification efficiency was influenced by the specific surface area of the artificial media, with the highest efficiency recorded for the LevaPor medium.

Another group of research papers published in the discussed Special Issue focused on fly ash from the incineration of municipal solid waste [61,62] or from coal combustion [63,64]. These papers assessed the properties of this type of waste or the possibilities of its practical use and evaluated the effects of its applications.

Fly ash from waste incineration is classified as hazardous waste. The direct disposal of such waste in landfills can not only result in the loss of valuable substitute materials but can also pose a potential risk of pollution to the aquatic environment due to the leaching of toxic components soluble in water. In paper [61], 12 fly ash samples from 9 municipal solid waste incineration (MSWI) plants in southeast China (collected from various types of grate furnaces and factory ash hoppers) were treated with chelating agents (dithioamine polymer and agents primarily targeting heavy metals) and analyzed for the content of polychlorinated dibenzo-p-dioxins/furans (PCDD/Fs) and heavy metals, using Chinese national and eco-environmental standards. The results were compared to previous fly ash data and Chinese landfill standards. To the best of the authors' knowledge, this was the first study to comprehensively analyze fly ash after treatment with chelating agents. The study presented and discussed the distribution of 17 of the most toxic PCDD/F homologues and 10 groups of homologues, examining the relationships between PCDD/Fs homologues and their toxicity. Additionally, for some samples, the distribution of several heavy metals in the fly ash leachate was determined. The results indicated that the fly ash samples from the MSWI plants were well below the national standard for dioxin emissions for waste to be disposed of in landfills. The heavy metal content was generally lower than the levels obtained using the national standard chelating agent, except for one type of fly ash.

Fly ash from waste incineration can be used, among other applications, in the production of concrete as a partial substitute for cement or solidified in cement blocks for safe storage or disposal via landfill. The most effective approach is to achieve the immobilization of toxic components in the fly ash within the concrete, thereby enabling its use for industrial purposes. This not only helps save energy but also limits CO_2 emissions and reduces the consumption of natural resources. Paper [62] explores the effect of various amounts of fly ash on the mechanical properties of concrete made with Portland cement CEM I, including compressive and bending strength, as well as the results of leaching tests for contaminants from these concretes. The study found that a 4% fly ash content in the concrete mix is the threshold at which the mechanical properties of cement mortars are not negatively affected. Exceeding this limit caused a swelling effect during the binding process. The leachability tests confirmed that chlorides and heavy metals from the fly ash were almost completely immobilized by the C-S-H phase (with an immobilization degree exceeding 99%), while the leachability of sulfates was limited to 96–97% compared to their leachability from the fly ash itself.

An extensive review of the properties and the methods for utilizing fly ash from coal combustion was presented in paper [63]. This study also includes original research on the potential use of a mixture of this type of fly ash with Portland cement (in a 3:7 weight ratio) as a low-cost material for the treatment of benzene-contaminated soil. The mixture was tested at soil incorporation levels of 40%, 60%, and 80% by weight. The study analyzed variations in the benzene levels emitted during the chemical solidification process of the soil using the cement–fly ash mixture and evaluated the physical properties (compressive strength and capillary water absorption) of the resulting monoliths. The research showed that the mixture could reduce the volume of benzene released during the process by up to 36% (during the first phase) when applied at an 80% weight ratio. Mechanical tests confirmed a direct relationship between the compressive strength and the amount of binding material (cement–fly ash mixture). An increase in the dosage also led to a reduction in both capillary water imbibition and water absorption by immersion. In conclusion, the authors stated that the cement–fly ash mixture can be considered an inexpensive and effective material for treating soils contaminated with low levels of benzene (with a test dose of 0.5 mL of benzene per 200 g of raw soil).

In the final paper discussed [64], fly ash from coal combustion was used as a filler in composites of rigid polyurethane (PUR) foam. The research focused on the thermal, mechanical, and flammability properties of these composites, which were produced with a 10% share of waste in the composite's mass. The study considered three types of fly ash modifications: silanization with 1% and 2% silane concentration (in relation to the fly ash mass), sieving to obtain a fraction with particles smaller than 75 μm, and a combination of both processes, with an unmodified fly ash sample used as a reference. The results showed that PUR materials filled with fly ash silanized with 1% and 2% silane exhibited the best properties, including the lowest apparent densities, water absorption, and brittleness values. PUR foam with fly ash modified with 1% silane solution also demonstrated good flame-retardant properties. However, the best performance in terms of the lowest smoke emission, including CO_2 and CO, during combustion was observed in the foam containing sifted fly ash. The authors concluded that the use of fly ash from coal combustion in the production of PUR foam is an excellent solution for reducing the use of petrochemical products in polyurethane materials manufacturing and offers an effective method for waste disposal.

3. Summary and Future Directions

The research papers presented in this Special Issue address several key challenges related to waste management. These studies adopt both systemic and detailed approaches to planning and implementing waste management strategies. One of their primary goals is to recognize and minimize the environmental impact of waste processes. The papers introduce new assessment methods or enhance existing ones, which help to better identify these impacts and guide optimal decision-making. Additionally, the experimental works presented contribute to the development of waste processing and utilization technologies. These innovations align with the principles of a circular economy, promoting raw material and energy savings while simultaneously improving environmental protection.

It has been confirmed that the Strategic Environmental Assessment (SEA) is a valuable tool for guiding the strategic planning process toward sustainable development goals, also in the field of waste management, where it supports the implementation of sustainable national waste management programs [54]. The established goals and indicators enable the assessment of the complex implications of this type of program and its interactions with the environment. Future research on the development and application of SEA should focus on combining various qualitative-expert and quantitative methods, including the use of semi-quantitative multi-criteria evaluation approaches [54].

One such method, replacing intuitive evaluations or expert opinions, is multicriteria analysis. This approach allows for a quantitative, multifaceted, and objective evaluation of alternative solutions, as demonstrated in the optimization of urban street cleaning scenarios to minimize chloride levels in sweepings [55]. Thus, multicriteria analysis can be effectively applied to complex decision-making tasks that involve multiple, often conflicting goals and socio-economic interests.

Complex data collected during the monitoring of waste management facilities, such as landfills, provide valuable insights that help permitting and regulatory authorities improve operational programs and action plans in compliance with applicable legal regulations. This can be achieved through the synoptic risk assessment method developed in [56], which enables the classification of landfills based on their environmental impact and sets priorities for necessary actions. Future research could focus on the statistical evaluation of key parameters to track trends in environmental impacts and assess whether conditions at individual landfills are improving, deteriorating, or stagnating. In addition to direct measurements, the behavior of substances released into the environment can be evaluated using modeling methods. For challenges related to water chemistry, mass and energy transport, secondary mineral precipitation, or the exploitation of geothermal deposits, specialized hydrogeochemical modeling systems such as PHREEQC or similar models can be employed [57]. Simulations using these systems can help identify trends and factors influencing mineral solubility changes, which would be difficult to predict without such modeling.

Important research methods in waste management systems include studies conducted in real facilities or laboratory experimental stations. These studies may aim to assess the effectiveness of current solutions, test new waste processing methods, or identify additional environmental impacts. The results of such studies can provide valuable insights that help operators optimize facility operations and assist decision-makers in selecting the most effective waste management strategies. For example, the optimization of refuse-derived fuel production in mechanical–biological treatment (MBT) plants can be achieved by adjusting key parameters of the biological drying (biodrying) process, such as bed temperature, and by screening the final products to maximize net heating value and minimize loss on ignition [58].

The use of compost produced from a mixture of sewage sludge with minimal structural material input (e.g., straw), which results in low initial C/N values, does not negatively affect the rate of biochemical changes. Moreover, the risk of Cu and Zn release from the waste can be effectively minimized by reducing the share of mobile heavy metal fractions and increasing their proportion in stable fractions [59]. For water purification in recirculating aquaculture systems that ensure the safety of high-nutrient fish cultures (e.g., certain species of salmonids), high-performance biofilters with passive bed activation can be utilized [60]. However, further research is recommended to confirm the effectiveness of this approach for different fish species and under varying technological conditions.

Fly ash from the incineration of municipal solid waste may contain harmful components, such as toxic heavy metals, polychlorinated dibenzo-p-dioxins, dibenzofurans, chlorides, sulfates, and others, which hinder its use or disposal in landfills [61,62]. One safe method for managing such ash is its use as a partial replacement for cement in concrete production. This not only immobilizes toxic components but also allows for the creation of new products with minimal environmental risk. This approach is environmentally neutral, economically viable, and aligns with the principles of the circular economy. To reduce the leaching of harmful substances from concrete, the addition of nanosilica can be effective. However, due to current cement standards, fly ash from municipal waste incineration can only be used in small amounts (about 4%) for special, chemically resistant cements that are not commercially available [61]. Further research is recommended to enhance the properties of concrete made with this type of fly ash, such as studies on chemical degassing of the ash to reduce its effects on swelling and cracking in cement mortars. Additionally, exploring the effectiveness of immobilizing toxic components by incorporating fly ash into geopolymer materials should also be considered.

The properties of fly ash from coal combustion are closely tied to the type of coal burned and the combustion process. Due to its chemical composition, the potential use of fly ash in agriculture is significantly limited. However, it can be utilized in the production of building materials, such as cement, cellular concrete, lightweight aggregates, ceramic tiles, or bricks, and as a sorbent material for removing metals and certain organic contaminants [63]. One promising application is using fly ash as a low-cost adsorbent, replacing materials like activated carbon, for adsorbing BTEX compounds and reducing their migration in soils contaminated with petroleum products or chemicals. A mixture of fly ash and Portland cement has been confirmed to be effective in the chemical solidification of soils contaminated with low levels of benzene. However, factors such as temperature conditions and the properties of the soil and fly ash can limit the effectiveness of this process [63]. This method may be particularly useful in developing countries that lack proper infrastructure for hazardous waste treatment. Another innovative application of fly ash is in the manufacturing of polyurethane (PUR) materials, such as PUR foam, where it acts as an additive (filler) to reduce the amount of petrochemical products used. Certain modifications can even improve the properties of fly ash–PUR composites. However, it is important to carefully select the material properties to be enhanced and adjust the modification method accordingly [64].

Author Contributions: Conceptualization, R.O. and K.G.; formal analysis, R.O.; writing—original draft preparation, R.O.; writing—review and editing, K.G.; supervision, R.O. and K.G.; project administration, R.O. and K.G. All authors have read and agreed to the published version of the manuscript.

Conflicts of Interest: The authors declare no conflicts of interest.

References

1. Brunner, P.H.; Rechberger, H. Waste to energy—key element for sustainable waste management. *Waste Manag.* **2015**, *37*, 3–12. [CrossRef] [PubMed]
2. Dri, M.; Canfora, P.; Antonopoulos, I.S.; Gaudillat, P. *Best Environmental Management Practice for the Waste Management Sector*; JRC Science for Policy Report; Publications Office of the European Union: Luxembourg, 2018.
3. Kasemsap, K. Environmental management and waste management: Principles and applications. In *Ethics and Sustainability in Global Supply Chain Management*; Akkucuk, U., Ed.; IGI Global: Hershey, PA, USA, 2017; pp. 26–49. [CrossRef]
4. Elsaid, S.; Aghezzaf, E.-H. A framework for sustainable waste management: Challenges and opportunities. *Manag. Res. Rev.* **2015**, *38*, 1086–1097. [CrossRef]
5. Tomić, T.; Schneider, D.R. Circular economy in waste management—Socio-economic effect of changes in waste management system structure. *J. Environ. Manag.* **2020**, *267*, 110564. [CrossRef] [PubMed]
6. Kalmykova, Y.; Sadagopan, M.; Rosado, L. Circular economy—From review of theories and practices to development of implementation tools. *Resour. Conserv. Recycl.* **2018**, *135*, 190–201. [CrossRef]
7. Grzesik, K.; Kossakowska, K.; Bieda, B.; Kozakiewicz, R. Screening Life Cycle Assessment of beneficiation processes for Rare Earth Elements recovery from secondary sources. *IOP Conf. Ser. Earth Environ. Sci.* **2019**, *214*, 012068. [CrossRef]
8. Sondh, S.; Upadhyay, D.S.; Patel, S.; Patel, R.N. Strategic approach towards sustainability by promoting circular economy-based municipal solid waste management system—A review. *Sustain. Chem. Pharm.* **2024**, *37*, 101337. [CrossRef]
9. Okorie, O.; Salonitis, K.; Charnley, F.; Moreno, M.; Turner, C.; Tiwari, A. Digitisation and the Circular Economy: A Review of Current Research and Future Trends. *Energies* **2018**, *11*, 3009. [CrossRef]
10. Rejeb, A.; Suhaiza, Z.; Rejeb, K.; Seuring, S.; Treiblmaier, H. The Internet of Things and the circular economy: A systematic literature review and research agenda. *J. Clean. Prod.* **2022**, *350*, 131439. [CrossRef]
11. Gregson, N. *The Waste of the World: Consumption, Economies and the Making of the Global Waste Problem*; Bristol University Press: Bristol, UK, 2023. [CrossRef]
12. Kůdela, J.; Smejkalová, V.; Šomplák, R.; Nevrlý, V. Legislation-induced planning of waste processing infrastructure: A case study of the Czech Republic. *Renew. Sustain. Energy Rev.* **2020**, *132*, 110058. [CrossRef]
13. Maheshwari, S.; Deswal, S. Role of waste management at landfills in sustainable waste management. *Int. J. Emerg. Technol.* **2017**, *8*, 324–328.
14. Merrild, H.; Larsen, A.W.; Christensen, T.H. Assessing recycling versus incineration of key materials in municipal waste: The importance of efficient energy recovery and transport distances. *Waste Manag.* **2012**, *32*, 1009–1018. [CrossRef] [PubMed]
15. Iravanian, A.; Ravari, S.O. Types of Contamination in Landfills and Effects on The Environment: A Review Study. *IOP Conf. Ser. Earth Environ. Sci.* **2020**, *614*, 012083. [CrossRef]
16. Morita, A.K.; Ibelli-Bianco, C.; Anache, J.A.; Coutinho, J.V.; Pelinson, N.S.; Nobrega, J.; Rosalem, L.M.P.; Leite, C.M.C.; Niviadonski, L.M.; Manastella, C.; et al. Pollution threat to water and soil quality by dumpsites and non-sanitary landfills in Brazil: A review. *Waste Manag.* **2021**, *131*, 163–176. [CrossRef]
17. Vaverková, M.D. Landfill Impacts on the Environment—Review. *Geosciences* **2019**, *9*, 431. [CrossRef]
18. Huang, D.; Du, Y.; Xu, Q.; Ko, J.H. Quantification and control of gaseous emissions from solid waste landfill surfaces. *J. Environ. Manag.* **2022**, *302*, 114001. [CrossRef] [PubMed]
19. Dabrowska, D.; Rykala, W.; Nourani, V. Causes, Types and Consequences of Municipal Waste Landfill Fires—Literature Review. *Sustainability* **2023**, *15*, 5713. [CrossRef]
20. Juan, W.Y.; Wu, C.L.; Liu, F.W.; Chen, W.S. Fires in Waste Treatment Facilities: Challenges and Solutions from a Fire Investigation Perspective. *Sustainability* **2023**, *15*, 9756. [CrossRef]
21. Grzesik, K.; Kossakowska, K.; Kozakiewicz, R.; Kowalewski, Z.; Gorzelnik, T.; Oleniacz, R.; Drzewiecki, W. Waste fires –waste of energy, waste of materials. In *New insights into Environmental and Energy Development*; Sornek, K., Ed.; Institute of Sustainable Energy: Kraków, Poland, 2021; pp. 5–11.

22. Mikalsen, R.F.; Lönnermark, A.; Glansberg, K.; McNamee, M.; Storesund, K. Fires in waste facilities: Challenges and solutions from a Scandinavian perspective. *Fire Saf. J.* **2021**, *120*, 103023. [CrossRef]
23. Oleniacz, R.; Drzewiecki, W.; Gorzelnik, T.; Grzesik, K.; Kozakiewicz, R.; Kowalewski, Z.; Kossakowska, K. Assessment of the impact of waste fires on air quality and atmospheric aerosol optical depth: A case study in Poland. *Energy Rep.* **2023**, *9*, 16–38. [CrossRef]
24. Dutta, A.; Patra, A.; Hazra, K.K.; Nath, C.P.; Kumar, N.; Rakshit, A. A state of the art review in crop residue burning in India: Previous knowledge, present circumstances and future strategies. *Environ. Chall.* **2022**, *8*, 100581. [CrossRef]
25. Jouhara, H.; Czajczyńska, D.; Ghazal, H.; Krzyżyńska, R.; Anguilano, L.; Reynolds, A.J.; Spencer, N. Municipal waste management systems for domestic use. *Energy* **2017**, *139*, 485–506. [CrossRef]
26. Koul, B.; Yakoob, M.; Shah, M.P. Agricultural waste management strategies for environmental sustainability. *Environ. Res.* **2022**, *206*, 112285. [CrossRef] [PubMed]
27. Ramadan, B.S.; Rachman, I.; Ikhlas, N.; Kurniawan, S.B.; Miftahadi, M.F.; Matsumoto, T. A comprehensive review of domestic-open waste burning: Recent trends, methodology comparison, and factors assessment. *J. Mater. Cycles Waste. Manag.* **2022**, *24*, 1633–1647. [CrossRef]
28. Kováts, N.; Hubai, K.; Sainnokhoi, T.A.; Eck-Varanka, B.; Hoffer, A.; Tóth, Á.; Kakasi, B.; Teke, G. Ecotoxic emissions generated by illegal burning of household waste. *Chemosphere* **2022**, *298*, 134263. [CrossRef] [PubMed]
29. Mentes, D.; Jordán, A.; Farkas, L.; Muránszky, G.; Fiser, B.; Viskolcz, B.; Póliska, C. Evaluating emissions and air quality implications of residential waste incineration. *Sci. Rep.* **2024**, *14*, 21314. [CrossRef]
30. Oleniacz, R.; Tomkowicz, A. Uncontrolled combustion of household wastes and vegetation remains in a rural community–questionnaire results. *Geomat. Environ. Eng.* **2009**, *3*, 71–79.
31. Aparcana, S. Approaches to formalization of the informal waste sector into municipal solid waste management systems in low-and middle-income countries: Review of barriers and success factors. *Waste Manag.* **2017**, *61*, 593–607. [CrossRef] [PubMed]
32. Ginevičius, R. The efficiency of municipal waste management systems in the environmental context in the countries of the European Union. *J. Int. Stud.* **2022**, *15*, 63–79. [CrossRef]
33. Fehr, A.; Urushadze, T.; Zöller, N.; Knerr, B.; Ploeger, A.; Vogtmann, H. Establishing a Sustainable Waste Management System in a Transitional Economic Context: Analysis of the Socio-Economic Dynamics. *Sustainability* **2020**, *12*, 3887. [CrossRef]
34. Kristensen, H.S.; Mosgaard, M.A.; Remmen, A. Integrating circular principles in environmental management systems. *J. Clean. Prod.* **2021**, *286*, 125485. [CrossRef]
35. Zorpas, A.A. Strategy development in the framework of waste management. *Sci. Total Environ.* **2020**, *716*, 137088. [CrossRef] [PubMed]
36. Campitelli, A.; Schebek, L. How is the performance of waste management systems assessed globally? A systematic review. *J. Clean. Prod.* **2020**, *272*, 122986. [CrossRef]
37. Allesch, A.; Brunner, P.H. Assessment methods for solid waste management: A literature review. *Waste Manag. Res.* **2014**, *32*, 461–473. [CrossRef] [PubMed]
38. Allesch, A.; Brunner, P.H. Material flow analysis as a decision support tool for waste management: A literature review. *J. Ind. Ecol.* **2015**, *19*, 753–764. [CrossRef]
39. Cano-Londoño, N.; Heriberto, C.; Krisztián, B. Integrated sustainability assessment: Exergy, emergy, life cycle assessment. *Front. Sustain.* **2022**, *3*, 921874. [CrossRef]
40. Sciubba, E.; Ulgiati, S. Emergy and exergy analyses: Complementary methods or irreducible ideological options? *Energy* **2005**, *30*, 1953–1988. [CrossRef]
41. Soltanian, S.; Kalogirou, S.A.; Ranjbari, M.; Amiri, H.; Mahian, O.; Khoshnevisan, B.; Jafary, T.; Nizami, A.S.; Gupta, V.K.; Aghaei, S.; et al. Exergetic sustainability analysis of municipal solid waste treatment systems: A systematic critical review. *Renew. Sustain. Energy Rev.* **2022**, *156*, 111975. [CrossRef]
42. Del Borghi, A.; Gallo, M.; Del Borghi, M. A survey of life cycle approaches in waste management. *Int. J. Life Cycle Assess.* **2009**, *14*, 597–610. [CrossRef]
43. Grzesik, K.; Malinowski, M. Life cycle assessment of refuse-derived fuel production from mixed municipal waste. *Energ. Sources Part A* **2016**, *38*, 3150–3157. [CrossRef]
44. Grzesik, K.; Malinowski, M. Life cycle assessment of mechanical–biological treatment of mixed municipal waste. *Environ. Eng. Sci.* **2017**, *34*, 207–220. [CrossRef]
45. Zhang, J.; Qin, Q.; Li, G.; Tseng, C.H. Sustainable municipal waste management strategies through life cycle assessment method: A review. *J. Environ. Manag.* **2021**, *287*, 112238. [CrossRef] [PubMed]
46. Ibáñez-Forés, V.; Bovea, M.D.; Coutinho-Nóbrega, C.; de Medeiros, H.R. Assessing the social performance of municipal solid waste management systems in developing countries: Proposal of indicators and a case study. *Ecol. Indic.* **2019**, *98*, 164–178. [CrossRef]

47. Khudyakova, T.; Lyaskovskaya, E. Improving the Sustainability of Regional Development in the Context of Waste Management. *Sustainability* **2021**, *13*, 1755. [CrossRef]
48. Salhofer, S.; Wassermann, G.; Binner, E. Strategic environmental assessment as an approach to assess waste management systems. Experiences from an Austrian case study. *Environ. Model. Softw.* **2007**, *22*, 610–618. [CrossRef]
49. Oleniacz, R. Assessment of the impact of municipal waste incineration plants on air quality and the possibilities of its reduction. *Pol. J. Environ. Stud.* **2014**, *23*, 95–104.
50. Yilmaz, O.; Kara, B.Y.; Yetis, U. Hazardous waste management system design under population and environmental impact considerations. *J. Environ. Manag.* **2017**, *203*, 720–731. [CrossRef] [PubMed]
51. Kasina, M.; Jarosz, K. Chemical and Microbiological Techniques for Recovery and/or Removal of Elements from Incinerated Sewage Sludge Ash—A Review of Basic Methods. *Energies* **2023**, *16*, 2840. [CrossRef]
52. Mnif, W.; Ben Rebah, F. Bioflocculants as Alternative to Synthetic Polymers to Enhance Wastewater Sludge Dewaterability: A Review. *Energies* **2023**, *16*, 3392. [CrossRef]
53. Atilgan, A.; Krakowiak-Bal, A.; Ertop, H.; Saltuk, B.; Malinowski, M. The Energy Potential of Waste from Banana Production: A Case Study of the Mediterranean Region. *Energies* **2023**, *16*, 5244. [CrossRef]
54. Josimović, B.; Manić, B.; Krunić, N. Strategic Environmental Assessment as a Support in a Sustainable National Waste Management Program—European Experience in Serbia. *Energies* **2022**, *15*, 4568. [CrossRef]
55. Gronba-Chyła, A.; Generowicz, A.; Kwaśnicki, P.; Cycoń, D.; Kwaśny, J.; Grąz, K.; Gaska, K.; Ciuła, J. Determining the Effectiveness of Street Cleaning with the Use of Decision Analysis and Research on the Reduction in Chloride in Waste. *Energies* **2022**, *15*, 3538. [CrossRef]
56. Cerar, S.; Serianz, L.; Koren, K.; Prestor, J.; Mali, N. Synoptic Risk Assessment of Groundwater Contamination from Landfills. *Energies* **2022**, *15*, 5150. [CrossRef]
57. Jakubowicz, P.; Steliga, T.; Wojtowicz, K. Analysis of Temperature Influence on Precipitation of Secondary Sediments during Water Injection into an Absorptive Well. *Energies* **2022**, *15*, 9130. [CrossRef]
58. Latosińska, J.; Żygadło, M.; Dębicka, M. The Biological Drying of Municipal Waste in an Industrial Reactor—A Case Study. *Energies* **2022**, *15*, 1039. [CrossRef]
59. Leśniańska, A.; Janowska, B.; Sidełko, R. Immobilization of Zn and Cu in Conditions of Reduced C/N Ratio during Sewage Sludge Composting Process. *Energies* **2022**, *15*, 4507. [CrossRef]
60. Nędzarek, A.; Bonisławska, M.; Tórz, A.; Tański, A.; Formicki, K. Effect of Filter Medium on Water Quality during Passive Biofilter Activation in a Recirculating Aquaculture System for Oncorhynchus mykiss. *Energies* **2022**, *15*, 6890. [CrossRef]
61. Pan, S.; Yao, Q.; Cai, W.; Peng, Y.; Luo, Y.; Wang, Z.; Jiang, C.; Li, X.; Lu, S. Characterization of Dioxins and Heavy Metals in Chelated Fly Ash. *Energies* **2022**, *15*, 4868. [CrossRef]
62. Czop, M.; Łaźniewska-Piekarczyk, B.; Kajda-Szcześniak, M. Evaluation of the Immobilization of Fly Ash from the Incineration of Municipal Waste in Cement Mortar Incorporating Nanomaterials—A Case Study. *Energies* **2022**, *15*, 9050. [CrossRef]
63. Banaszkiewicz, K.; Marcinkowski, T.; Pasiecznik, I. Fly Ash as an Ingredient in the Contaminated Soil Stabilization Process. *Energies* **2022**, *15*, 565. [CrossRef]
64. Zygmunt-Kowalska, B.; Pielichowska, K.; Trestka, P.; Ziąbka, M.; Kuźnia, M. The Effect of Ash Silanization on the Selected Properties of Rigid Polyurethane Foam/Coal Fly Ash Composites. *Energies* **2022**, *15*, 2014. [CrossRef]

Disclaimer/Publisher's Note: The statements, opinions and data contained in all publications are solely those of the individual author(s) and contributor(s) and not of MDPI and/or the editor(s). MDPI and/or the editor(s) disclaim responsibility for any injury to people or property resulting from any ideas, methods, instructions or products referred to in the content.

Article

Strategic Environmental Assessment as a Support in a Sustainable National Waste Management Program—European Experience in Serbia

Boško Josimović *, Božidar Manić and Nikola Krunić

Institute of Architecture and Urban and Spatial Planning of Serbia, 11000 Belgrade, Serbia; bozam@iaus.ac.rs (B.M.); nikola@iaus.ac.rs (N.K.)
* Correspondence: bosko@iaus.ac.rs

Citation: Josimović, B.; Manić, B.; Krunić, N. Strategic Environmental Assessment as a Support in a Sustainable National Waste Management Program—European Experience in Serbia. *Energies* 2022, 15, 4568. https://doi.org/10.3390/en15134568

Academic Editors: Robert Oleniacz and Katarzyna Grzesik

Received: 8 June 2022
Accepted: 21 June 2022
Published: 22 June 2022

Publisher's Note: MDPI stays neutral with regard to jurisdictional claims in published maps and institutional affiliations.

Copyright: © 2022 by the authors. Licensee MDPI, Basel, Switzerland. This article is an open access article distributed under the terms and conditions of the Creative Commons Attribution (CC BY) license (https://creativecommons.org/licenses/by/4.0/).

Abstract: Strategic Environmental Assessment (SEA) is one of the most important instruments for directing the strategic planning process toward the sustainable development goals in various areas of human activity. This also applies to the field of waste management. By applying SEA in waste management planning, it is possible to see the benefits and consequences of the proposed changes in space that will occur during the implementation of strategic planning concepts and based on that make appropriate decisions respecting the capacity of the space where the planned activities are implemented. The paper presents the application of SEA for the National Waste Management Program with all its spatial, organizational, energy, environmental, and other solutions, and the way they are included in the specific method of multicriteria evaluation in SEA. The specificity of the methodological approach indicates the need for equal consideration of environmental and socio-economic aspects of development and a clear presentation of the results obtained in order to make optimal decisions in waste management planning at the national level. The National Waste Management Program in Serbia, which legislation in the field of environmental protection and waste management is harmonized with EU legislation and directives, was chosen for the case study. The obtained results indicate the importance of an interdisciplinary approach in the evaluation of strategic solutions in the field of waste management, which is achieved by specific choice of environmental and socio-economic SEA goals and indicators as a basis for valorization of the proposed concept of waste management.

Keywords: waste management; strategic environmental assessment; strategic planning; multicriteria evaluation; decision making

1. Introduction

The term Strategic Environmental Assessment (SEA) was launched in 1989 in the UK, with the concept itself derived from project-oriented environmental impact assessment (EIA). The principles of SEA and EIA were the same [1]. Over time, the range of SEA interpretations has begun to expand and was used for other types of assessments that differ from those based on the principles of project EIA. Recent definitions describe the SEA less rigorously, as "a systematic process for evaluating the environmental impact of a proposed policy, plan or program to ensure that it is fully integrated and adequately addressed at the earliest possible stage of decision-making, on an equal footing with economic and social considerations" [2]. International financial institutions, such as the World Bank, consider the SEA to be "a participatory approach to increasing the impact of social and environmental issues on development planning, decision-making, and implementation processes at the strategic level" [3].

A large number of authors have written about the role and importance of SEA in policy-making in various spheres of social action and the role of this instrument in decision-making since the early 1990s [4–7]. Additionally, the European Strategic Environmental

Assessment Directive 2001/42/EC stipulates that the SEA is carried out for plans, programs, and foundations in various areas, inter alia, the waste management area. By applying SEA in waste management planning, it is now possible to see the consequences of proposed planning solutions and changes in space, taking into account the needs of the space and the environment users and defining adequate measures for protection and monitoring of potentially endangered elements of the environment, with the inevitable involvement of the public in all phases of the development and adoption of the SEA. In this context, it is evident that SEA contributes to the decision-making process in waste management planning [7–13].

SEA represents a contribution to the evaluation of the key strategic planning solutions at the national and/or regional strategic level of waste management [14]. The paper elaborates on the implementation of the SEA for the National Waste Management Program in the Republic of Serbia for the period 2022–2031 (NWMP). NWMP is the highest strategic document at the national level in the field of waste management in Serbia. The results of the SEA process provide support in making decisions on the eligibility conditions of the NWMP in relation to environmental objectives. In the context of good decision making for obtaining the sustainability of solutions defined in NWPM, consideration of various aspects of possible impacts is based on a holistic approach that unequivocally points to the application of multicriteria evaluation in the development of SEA for waste management plans [15–23]. The SEA concept for NWMP was based on this specific and innovative approach to a modern waste management system, which is elaborated below. The SEA was implemented for the needs of the NWMP, which sets strategic goals for improving the waste management system and the basic principles that should guide all waste management participants to achieve these goals in Serbia for the period from 2022 to 2031. The implementation of this Program, in addition to reducing the harmful impact on the environment and climate change, should enable the realization of preconditions for the use of waste in the circular economy, development goals, and measures determined in a special program. Additionally, special programs are being developed for the establishment of a waste sludge management system from wastewater treatment plants and for the treatment of animal waste. Management of agricultural, mining, and medical and pharmaceutical waste is planned by sectoral planning documents. The vision resulting from the analysis of the current situation and potential in the field of waste management is to minimize the impact of waste on the environment and increase resource efficiency on the principles of circular economy, which provides control of waste generation, waste utilization, and incentives to invest and affirm economic opportunities. This vision can be achieved by consistently applying principles based on reducing environmental pressures and ensuring a better quality of life for citizens, maintaining a clear and sustainable development perspective, and building a supportive environment for the establishment of a circular economy model. Following the vision, the general and specific goals of waste management have been determined and the measures and instruments, and activities necessary for their realization, have been developed. The overall goal is to develop a sustainable waste management system to conserve resources and reduce negative environmental impacts and space degradation. This includes reducing the amount of generated waste, reducing the amount of waste in landfills that can be reused as raw material, energy source, or in some other way, reducing the share of biodegradable waste in municipal waste, reducing the negative impact of landfilled waste on the environment, climate and human health, and waste management according to the principles of the circular economy.

To achieve the general goal of the Program, special goals have been set:

- Improved municipal waste management system through increased recycling rate and reduced waste disposal in the unsanitary landfills;
- Established system of sustainable management of hazardous and industrial waste;
- Increased rate of collection, reuse, and recycling of special waste streams and more efficient usage of resources;

- Strengthened capacity of institutions in the field of waste management and harmonized regulations with EU regulations.

Along with the Law on Waste Management and the Regulation on Landfills, the NWMP is the main strategic document for the establishment and implementation of waste management systems in Serbia. The development of the SEA for NWMP was approached to direct the planning process toward the goals of sustainable development, i.e., achieving the goals defined in the SEA related to environmental protection and socio-economic aspects of development. The obtained results were the basis for deciding on the sustainability of the NWMP.

The originality of the scientific approach is reflected in the specific choice of SEA goals and indicators in relation to which the procedure of multi-criteria evaluation of variant and software solutions is carried out. An interdisciplinary approach is, therefore, realized in the assessment of the anticipated changes in space and environment that are expected during the implementation of the NWMP. In addition, applying this approach creates assumptions that decision-makers have a clear insight into the expected results (positive impacts) and implications/consequences (negative impacts) of the proposed changes, in relation to the symbiosis of environmental and socio-economic aspects of development.

This paper is designed as follows: after the introductory part (Section 1), which is a review of literature and research in the field of SEA implementation, presents the methodological framework used in the SEA for NWMP (Section 2), and then summarizes the process of evaluating software solutions in NWMP—case study (Section 3). After that, in Section 4, the results of the conducted procedure are elaborated through a discussion, and the advantages, problems, and proposed directions of further research in the field of SEA application are given in the conclusion (Section 5).

2. Methodological Framework

Theoretical studies in the field of environmental and waste management planning [24–26] aim to directly define appropriate waste management systems and waste planning methods. The methodological frameworks used in the SEA process itself appear to be an important tool for planning a sustainable waste management system [7]. However, the concept of SEA methodologies, unlike the diverse precise software and highly operable tools used in environmental engineering or other science-based fields, is rather vague [27]. By analyzing the theoretical assumptions about the possibility of applying appropriate methodologies and scientific methods in the SEA procedure [28–33], we can conclude that SEA relies on qualitative consideration and techniques, and therefore expert assessment plays a more important role. The methodological framework in the development of the SEA is therefore based on a planning approach and the application of multicriteria evaluation of planned strategic determinants in relation to the capacity of space as a basis for the valorization of space for sustainable development. The procedure and methodological framework of the SEA are presented in Figure 1.

As shown in Figure 1, the initial stage in the SEA process is to decide on the need to develop and encompass the SEA, with the participation of the professional public and relevant institutions. After this follows the analytical part of the SEA which includes analysis of the plan document, i.e., planned concepts and strategic guidelines, and GIS tools-based analysis of the current state of the environment [34], analysis of relations with other planning documents and strategies, and identification of environmental problems. Based on the analytical work, the objectives of the SEA and the related indicators are determined, and the criteria for evaluation are defined. This is followed by an impact assessment procedure wherein the first phase the impact of variants/scenarios is assessed and then the most favorable variant is selected. Qualitative expert evaluation of variant solutions by NWMP sectors is performed in relation to the SEA objectives and relevant criteria. Then the process of multicriteria evaluation (semiquantitative method) begins, which is the spotlight of this paper. Based on the results of the multicriteria evaluation, actions are defined to limit the possible negative effects of NWMP in the process of its implementation. The SEA report summarizes all the results of the SEA (including opinions obtained through the public participation process) and makes decisions based on the non-adoption, necessary changes, or abandonment of the NWMP.

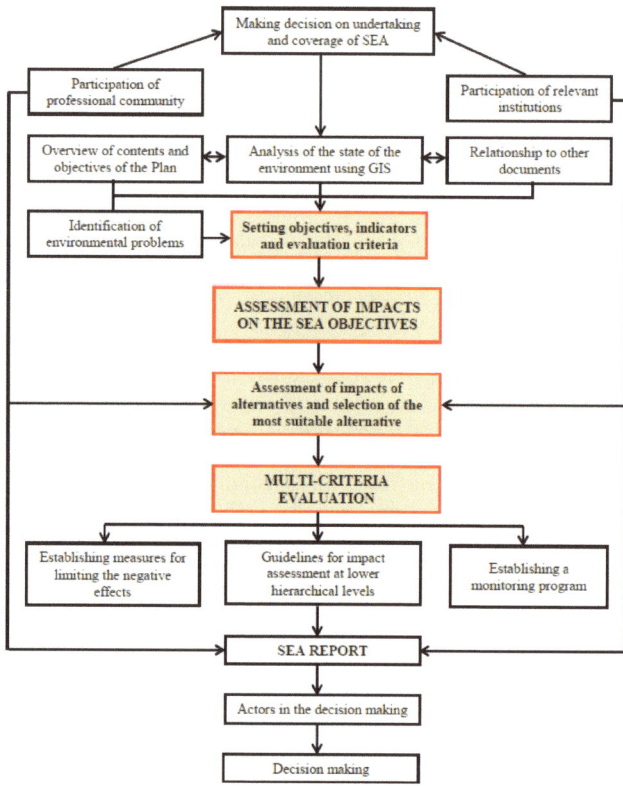

Figure 1. SEA Procedure and Methodological Framework.

3. Case Study: SEA for NWMP

The multicriteria evaluation of possible solutions, activities, and the NWMP conceived solutions is the key phase in the development of the SEA. It is performed for all NWMP strategic planning solutions concerning the defined SEA objectives and related indicators and also in relation to the evaluation criteria.

3.1. SEA Goals and Indicators

A sensitive step in this phase is to define goals and indicators. Defining the SEA objectives for the NWMP was conditioned by the results and predictions of the analytical work, and it resulted in the definition of 16 specific SEA objectives and 21 related indicators against which the sustainability of the solutions proposed within the NWMP was assessed. The goals are set in relation to environmental receptors and cover all aspects of sustainable development. The selection of SEA objectives and relevant indicators (Table 1) is harmonized with the Rulebook on the National List of Environmental Indicators [35]. The selection of indicators listed in Table 1 is in line with the planned activities in the field of implementation of the Program and its possible impacts on the quality of the environment, and will be used to evaluate the solution.

Table 1. Selection of SEA targets for relevant indicators in relation to environmental receptors.

SEA Area	General Objectives of SEA	Specific Objectives of SEA	Indicators
AIR AND CLIMATE CHANGE	Reduction of the air pollutants levels	-To reduce emissions of pollutants into the air -Introduction of cleaner technologies	-Reduction of air pollutants (%) -Change in greenhouse gas emissions, primarily CH_4 and CO_2 (%) -Introduction of BAT and increase of RES share in the balance (%)
WATER	Protection and preservation of surface and groundwater quality	-To reduce surface and groundwater pollution to levels that do not adversely affect the quality -To mitigate the negative impact of waste on the hydrological regime and water quality	-Serbian Water Quality Index (SWQI) * -Change of water regime -Contaminated (untreated) wastewater * -Change of water quality class (%)
LAND	Protection and sustainable use of forest and agricultural land	-Protection of forest and agricultural land -To reduce soil degradation -To minimize the area of land contaminated with waste management activities	-Change in forest land area (%) -Change in agricultural land area (%) -Area of land contaminated due to waste management activities (ha) -Area of land that has been rehabilitated
NATURAL VALUES	Protection, conservation, and improvement of landscapes, natural values and biodiversity, and geodiversity	-Landscape protection -Protection of natural values and areas -To preserve biodiversity and geodiversity—avoid irreversible losses	-Share of the recultivated in the total area of degraded areas (%) -Management of contaminated sites * -Number of endangered species of flora and fauna that may be affected by waste management activities
CULTURAL AND HISTORICAL HERITAGE	Preserve protected cultural assets	-Protection of cultural property, preservation of historical buildings and archaeological sites	-Number and importance of protected immovable cultural assets that may be affected by the activities of the waste management sector
SOCIO-ECONOMIC DEVELOPMENT	Improvement of the health of the population	-To reduce the impact of waste on the health of the population	-Frequency of diseases that can be associated with inadequate waste management -Number of people affected by the noise produced by a waste transport
	Strengthening of the institutional capacity	-To strengthen capacities for waste management -Developing awareness and public participation	-Measures to strengthen the capacity of the administration -Number of participatory programs
	Stimulation of the economic development	-To stimulate the economic development -To promote local employment	-Local government revenues from the waste management sector -Reduction of the number of unemployed as a result of employment in the waste management sector (%)

* definition, description of indicators, and calculation methodology are given in the Annex to the Rulebook on the National List of Environmental Indicators ("Official Gazette of RS", No. 37/11).

3.2. SEA Evaluation Criteria

Analyses of the possibilities of the implementation primarily of spatial, but also problem perception of possible impacts, resulted in the decision for the SEA to form 5 groups of criteria with a total of 18 individual criteria.

The first group of criteria was defined for monitoring environmental trends in different NWMP variants/scenarios. These are general criteria, adequate for this phase of the evaluation. Criteria for evaluating NWMP variant solutions are shown in Table 2.

Table 2. Criteria for evaluating variant solutions.

Label	Trend
+	overall positive impact
−	total negative impact
0	there is no direct influence or vague influence

The other 4 groups of criteria were used in the multi-criteria evaluation of planning solutions, and they relate to the size (intensity) of the impact; spatial dimensions (spatial dispersion) on which influence can be achieved; the likelihood that some estimated impact will occur in reality; and frequency (duration) of impacts (Table 3).

Table 3. Criteria for assessing the impact of individual strategic decisions from the NWMP.

Kind of the Impact	The Spatial Dimension of the Impact	Probability of the Impact	Frequency of Impact
Very favorable (+3) **Favorable (+2)** Positive (+1) Neutral (0) Negative (−1) **Unfavorable (−2)** Very negative (−3)	**National (N)** **Regional (R)** **Municipal (M)** Local (L)	Quite sure (Q) Likely (Lk) Possible (Ps) Unlikely (U)	Temporary (T) Long-term (Lt)

The significance of the identified impacts for achieving these goals is evaluated based on the criteria for estimating the size and spatial scale of the impact of planning solutions on the SEA goals. The NWMP's strategically important impacts are those that have strong or greater (positive or negative) effects at the national, regional, or municipal level (bolded criteria in Table 2).

3.3. Multicriteria Evaluation in SEA for NWMP

The phase of evaluation of plan variants and selection of the most favorable variant was the phase that preceded the semiquantitative method of multicriteria evaluation of individual solutions in the NWMP. This is the phase of monitoring in which environmental trends may arise as a consequence (negative trends) or as a result (positive trends) of the implementation of strategic solutions. Identification of positive and negative impacts of NWMP variants is performed by matrices in which variant solutions are crossed by NWMP sectors in relation to SEA objectives, and according to the criteria from Table 2 (Table 4). In this specific case, the SEA for NWMP processed the following variant solutions:

- variant A—reference scenario ("business as usual") which implies the continuation of the application of existing practices in waste management
- variant B—a scenario with the application of NWMP and all anticipated propositions

After assessing the impact of plan variants and predicting possible and positive trends in the environment that imply variant solutions of the plan, a decision was made on the selection of the most favorable variant solutions. The selection of the most favorable variant solutions represent the first significant contribution of SEA in the process of waste management planning, because in this phase the variants of the plan that may imply significant negative impacts on the environment and socio-economic aspects of development are eliminated.

After selecting the most favorable variant solutions, the selection of key and priority planning solutions is made, which will be included in the process of multicriteria evaluation (Table 5).

Table 4. Illustrative overview of the matrix for assessing the impact of variant solutions of the plan in relation to the objectives of the SEA for NWMP.

Area of Development	Variants	Development Scenario	SEA Goals															
			1	2	3	4	5	6	7	8	9	10	11	12	13	14	15	16
Establishing a sustainable waste management system	A	The establishment of a waste management system is not integral but partial and partially realized, which is not in line with modern principles and good practices. Problems are not solved systemically, which leads to several problems in the functioning of the system, and problems in the environment become more pronounced. Due to inadequate pre-treatment of waste at the regional level, increasing amounts of waste are generated, which are disposed of inadequately in local landfills. The dynamics of the (un)implemented proposition of the existing Strategy led to the fact that instead of building unique regional waste management centers with the necessary infrastructure, there are still a large number of local landfills which, due to their functioning, have negative impacts on environmental quality. All this implies problems in the environment and institutional organization, and waste is not used properly as a resource that is neither economically nor environmentally acceptable.	−	−	−	−	−	−	−	−	−	−	0	−	0	0	0	0
	B	Introduction of primary waste separation, reduction of waste production, an increase of waste recycling rate. Establishment of an integrated network of waste management facilities. Reduction of biodegradable waste disposed of in landfills. Establishment of a system for collecting special waste streams. Closure and remediation of existing local landfills and ban on disposal of untreated waste at the landfill. Extending the scope of the waste collection service. All these activities will contribute to all aspects of environmental protection. Changes in the way of financing the waste management system, establishing a market in waste management, and institutional and organizational strengthening of this area, altogether will achieve multiple positive effects on the environment and socio-economic aspects of development in the waste management sector.	+	+	+	+	+	+	+	+	+	+	+	+	+	+	+	+

Table 4. Cont.

Institutional changes	A	...	0	0	0	0	0	0	0	0	0	0	0	0	0	0	0	0
Institutional changes	B	...	0	0	0	0	0	0	0	0	0	0	0	0	+	+	+	+
Socio-economic aspects	A	...	−	−	−	−	−	−	−	−	−	−	−	−	0	−	−	0	−	−
Socio-economic aspects	B	...	+	+	+	+	+	+	+	+	+	0	+	+	+	+	+

Table 5. Solutions in NWMP covered by the multicriteria evaluation.

Num.	NWMP Program Solutions
1.	Reduction of waste production and increase of waste recycling rate following the EU directives
2.	Establishment of an integrated network of municipal waste management facilities
3.	Extension of waste collection coverage up to 100%
4.	Establishment of a network of transfer (transshipment) stations
5.	Establishment of a network of recycling yards
6.	Reduction of biodegradable waste in landfills with monitoring of the success of taken measures
7.	Biowaste composting
8.	Construction of a waste energy production plant in Belgrade
9.	Construction and fully functional and infrastructural equipment of regional sanitary landfills for non-hazardous waste
10.	Construction of regional hazardous waste warehouses
11	Closure and remediation of existing landfills and reclamation sites
12.	Establishment of mobile facilities for treatment of construction and demolition mineral waste
13.	Construction of a national facility for Physico-chemical treatment of hazardous waste
14.	Establishing capacities for incineration of organic industrial and medical waste at the national level
15.	Construction of a landfill/cassette for hazardous waste disposal
16.	Construction of large plants for biological treatment of bio-waste separated at the place of origin
17.	Construction of advanced RDF plants for mixed municipal households waste
18.	Waste management organization, including the division of responsibilities between the public and private sectors in the field of waste management
19.	Improving the institutional set-up for waste collection (inspection, training, establishment of regional companies, improvement of information system and reporting)
20.	Financing the waste management measures
21.	Conducting information campaigns on waste management for citizens
22.	Providing treatment for environmentally friendly waste in Serbia
23.	Strengthening the environmental inspection
24.	Measures to conduct public awareness campaigns and inform the general public or target groups of stakeholders
25.	Contaminated sites management
26.	Measures and instruments for the implementation of the Program

In this phase, and according to the similar principle as in the phase of the evaluation of variant solutions, matrices were formed as well (Tables 6 and 7). In them, all planning solutions shown in Table 5 intersect with the objectives of the SEA and are evaluated based on the first two groups of criteria from Table 3—significance/size of the impact and spatial scale/dispersion of the impact. Matrices were formed only for the first two groups of criteria because they were sufficient to identify strategically significant impacts, which is elaborated in point 3.2.

After the multicriteria evaluation of planning solutions, based on the results presented in the matrices (Tables 6 and 7), the identification of strategically significant impacts of planning solutions was approached by synthesizing the key impacts of the plan on the defined SEA objectives (Table 8).

Table 6. Illustrative presentation of the assessment of the importance of the impact of the NWMP program solutions.

Solution in NWMP	SEA Goals															
	1	2	3	4	5	6	7	8	9	10	11	12	13	14	15	16
Reduction of waste production and increase of waste recycling rate in accordance with EU directives	+1	0	+1	+1	+1	+1	+2	+2	+2	+1	0	0	0	0	+2	+1
Construction of regional hazardous waste warehouses	−3	0	−3	0	0	−3	−2	0	0	−3	0	−2	+3	0	0	+1
Measures and instruments for the implementation of the Program	+2	0	+2	+2	+2	+2	+2	+1	+1	+1	+1	0	+2	0	0	0

Table 7. Illustrative presentation of the spatial scale assessment of the NWMP program solutions.

Solution in NWMP	SEA Goals															
	1	2	3	4	5	6	7	8	9	10	11	12	13	14	15	16
Reduction of waste production and increase of waste recycling rate in accordance with EU directives	N		N	N	N	N	N	N	N	N					L	L
Construction of regional hazardous waste warehouses	L		R			L	L			L		L	N			L
Measures and instruments for the implementation of the Program	N		L	L	L	L	L	L	L	L	L		N			

Table 8. Illustrative presentation of the identification of strategically significant impacts of the NWMP program solutions.

Solution in NWMP	Identification of the Impact Rank		Explanation
	SEA Goal	Impact Rank *	
Reduction of waste production and increase of waste recycling rate in accordance with EU directives	7	N + 2/Lk/Lt	Greater, positive, long-term effects of national importance are expected in: the treatment of waste as a resource; reducing uncontrolled landfilling, i.e., reducing the amount of waste disposed of, meeting national, regional and local targets in the waste management sector, minimizing inadequate waste management, meeting national recycling targets, increasing investment in waste management system elements.
	8	N + 2/Lk/Lt	
	9	N + 2/Lk/Lt	
Construction of regional hazardous waste warehouses	1	L − 3/Ps/T	Strong negative impacts on basic environmental factors are possible in the case of hazardous waste ending up in the environment during transport or storage. There are some strong long-term positive effects of a national character in the context of strengthening the hazardous waste management capacity, which has not been adequately addressed at a national level so far.
	3	R − 3/Ps/T	
	6	L − 3/Ps/T	
	10	L − 3/Ps/T	
	13	N + 3/Q/Lt	
	15	N + 2/Ps/Lt	

Table 8. *Cont.*

Solution in NWMP	Identification of the Impact Rank		Explanation
	SEA Goal	Impact Rank *	
Measures and instruments for the implementation of the Program	1	N + 2/Lk/Lt	Measures and instruments for the implementation of the Program, which are structured through general waste management measures; hazardous waste management measures; construction and demolition waste management measures; and measures for the management of special waste streams, will have a greater positive impact on environmental factors and strengthening of the organizational, financial, and institutional capacity for waste management at the national level.
	13	N + 2/Q/Lt	

* Determining the rank of impact according to the criteria in Table 3.

The presentation of identified strategically significant impacts is tabular as in Table 8, where, in addition to determining the rank of impacts (column 3), an explanation of the impacts is given.

This completes the multicriteria evaluation process, which is the basis for defining appropriate measures to limit negative impacts, guidelines for impact assessments at lower hierarchical levels, and monitoring programs, which are also an integral part of the SEA process and SEA study. All of the above is the basis for making appropriate decisions about the NWMP.

4. Results and Discussion

The NWMP is a strategic framework for the implementation of waste management policies and measures at the national level. The possible implications that may arise in the environment as a result of the implementation of the NWMP and the significant participation of the public in the decision-making process unequivocally indicates the need for careful consideration of this aspect when designing waste management policy. The nature of the planned activities and possible impacts, on the one hand, and the significant spatial coverage of the NWMP, on the other, is the reason for the significant public interest in waste management plans. In this context, the role of environmental impact assessment is especially important. In the earliest phase of waste management policy development, it has the function of a control instrument directing the entire strategic planning process towards sustainability goals. It is the SEA that meets these specific requirements.

NWMP is specific since it conceives a substantially changed waste management system compared to the existing one, which was assessed in the SEA as unsustainable and environmentally unacceptable (variant A—business as usual). By applying the SEA process and multicriteria evaluation of variant and strategic solutions in the NWMP, all participants in the process of developing and adopting the SEA were able to see all space and environment-related key trends expected during the implementation of the NWMP.

Identification of the negative impacts of NWMP individual planning solutions, such as, e.g., NWMP program solution "Construction of regional hazardous waste warehouses" (Tables 6–8) and other solutions not presented in this paper, provides a basis for defining appropriate measures for environmental protection and monitoring, as well as for defining guidelines for impact assessments (SEA or EIA) at the lower hierarchical levels.

The motive for the analysis presented in this paper was to approach the concept of the application of the semi-quantitative method of multicriteria evaluation in SEA for NWMP. This would increase subjectivity in designing optimal solutions and conclusions in SEA and thus reduce the usual methodological shortcoming of the SEA—subjectivity of expert knowledge-dependent process and experiences. This, on the one hand, was partially achieved by forming the first group of criteria from Table 3, and, on the other hand, it gave a clearer idea of the directions of further research in the field of SEA methodology development. Further research referred to in the conclusion of this paper would achieve a significant methodological step forward in the development and implementation of SEA, which has just begun with this paper and the conclusions presented at its end. In the SEA for NWMP, it was relatively easy to suggest to decision-makers which solution is most favorable for NWMP implementation and which program solutions cause implications in space and the environment. This was achieved by a methodologically sound approach, a clear way of presenting the results that enabled extensive public participation in the critical phases of the SEA, and the use of the semi-quantitative method in impact assessment as an appropriate approach for a strategic document such as the NWMP. Although susceptibility to political decisions is inevitable and almost always threatens professional decisions, in this case, the SEA's propositions to the NWMP unequivocally refers to decisions that are in the interest of environmental protection, so decision-making was fully in line with SEA recommendations.

5. Conclusions

The specificity and advantages of the presented SEA approach are reflected in the identification of objectives and indicators of SEA, which is based on the analysis of the complex symbiosis of environmental quality, strategic frameworks defined in various strategic and planning documents, and NWMP. The objectives and indicators obtained by this approach represent a good basis for assessing the complex implications of NWMP in space and the environment and the possible interactions of different sectoral commitments on the elements of sustainable development. A clear matrix presentation of the obtained results is particularly suitable for presenting the results of the multicriteria evaluation, which is especially important in the SEA phases with the participation of the public. However, at the level of strategic planning and management, it is not necessary, and due to the lack of appropriate inputs, it is often not possible to use different mathematical methods, such as ARAS—Additive Ratio Assessment [36], AERMOD [37], or AHP—Analytical Hierarchy Process [38], or a holistic and inclusive approach that brings together different actors and disciplines for a successful transition to a circular economy [39]. The results of the assessment in the SEA represent the basis for establishing adequate guidelines when applying these methods and some other methods at a lower hierarchical level of impact assessment, i.e., when developing EIA—Environmental Impact Assessment and ESIA (Environmental Social Impact Assessment) [40]. Therefore, this shortcoming should be understood conditionally, but it should not be neglected in the process of drafting the SEA and making appropriate decisions. It is especially important to increase objectivity in the SEA process by using the above mentioned and other software packages and mathematical methods whenever the specifics of strategic documents allow it. In this context, further research in the development and application of SEA should be directed towards combining qualitative-expert-subjective methods with compatible and applicable quantitative methods (based primarily on GIS tools, but without excluding different software models for so-called 'partial' assessment of individual elements of the environment in the SEA process). In other words, the future of the SEA should be sought in the application of semi-quantitative methods of multicriteria evaluation. When it comes to public participation in the SEA process, attention should be paid to challenges arising in specific circumstances, such as those present during the period and conditions of the COVID-19 pandemic [41].

Author Contributions: Writing—original draft preparation, review and editing B.J.; validation and formal analysis, B.M.; visualization, N.K. All authors have read and agreed to the published version of the manuscript.

Funding: This paper is a result of research funded by the Ministry of Education, Science and Technological Development of the Republic of Serbia, contract number 451-03-9/2021-14/200006.

Institutional Review Board Statement: Not applicable.

Informed Consent Statement: Not applicable.

Data Availability Statement: The data presented in this study are publicly available in the SEA for the National Waste Management Program in the Republic of Serbia for the period 2022–2031 (in Serbian), accesible at https://www.ekologija.gov.rs/sites/default/files/inline-files/SEA%20 Program_otpad%20RS_0.pdf (accessed on 1 June 2022).

Conflicts of Interest: The authors declare no conflict of interest.

References

1. Fischer, T.B. Strategic environmental assessment in post-modern times. *Environ. Impact Assess. Rev.* **2003**, *23*, 155–170. [CrossRef]
2. Sadler, B.; Verheem, R. *Strategic Environmental Assessment: Status, Challenges and Future Directions*; Ministry of Housing, Spatial Planning and the Environment of The Netherlands: The Hague, The Netherlands, 1996; Available online: https://www.eia.nl/documenten/00000529.pdf (accessed on 1 June 2022).
3. Dalal-Clayton, B.; Sadler, B. *Strategic Environmental Assessment: A Sourcebook and Reference Guide to International Experience*; Earthscan: London, UK, 2005.
4. Nilsson, M.; Björklund, A.; Finnveden, G.; Johansson, J. Testing a SEA methodology for the energy sector: A waste incineration tax proposal. *Environ. Impact Assess. Rev.* **2005**, *25*, 1–32. [CrossRef]
5. Maričić, T.; Josimović, B. Overview of strategic environmental assessment (SEA) systems in SEE countries. *Arhit. I Urban.* **2005**, *16–17*, 66–74.
6. White, L.; Noble, B.F. Strategic environmental assessment for sustainability: A review of a decade of academic research. *Environ. Impact Assess. Rev.* **2013**, *42*, 60–66. [CrossRef]
7. Salhofer, S.; Wassermann, G.; Binner, E. Strategic environmental assessment as an approach to assess waste management systems. Experiances from an Austrian case study. *Environ. Model. Softw.* **2007**, *22*, 610–618. [CrossRef]
8. Arbter, K. SEA of Waste Management Plans–An Austrian Case Study. In *Implementing Strategic Environmental Assessment. Environmental Protection in the European Union*; Schmidt, M., João, E., Albrecht, E., Eds.; Springer: Berlin/Heidelberg, Germany, 2005; pp. 621–630. [CrossRef]
9. Desmond, M. Identification and development of waste management alternatives for Strategic Environmental Assessment (SEA). *Environ. Impact Assess. Rev.* **2009**, *29*, 51–59. [CrossRef]
10. Josimović, B.; Marić, I. Methodology for the Regional Landfill Site Selection. In *Sustainable Development–Authoritative and Leading Edge Content for Environmental Management*; Curkovic, S., Ed.; IntechOpen: London, UK, 2012; pp. 513–538. [CrossRef]
11. Josimović, B.; Ilić, M.; Filipović, D. *Planiranje Upravljanja Komunalnim Otpadom [Planning of Municipal Waste Management]*; IAUS: Belgrade, Serbia, 2009; Available online: https://raumplan.iaus.ac.rs/handle/123456789/543 (accessed on 1 June 2022).
12. Nenković-Riznić, M.; Josimović, B.; Milijić, S. SEA as instrument in responsible planning of tourist destinations. case study of Djerdap National Park, Serbia. *J. Environ. Tour. Anal.* **2014**, *2*, 5–18.
13. Josimović, B. *Planiranje Prostora u Sistemu Upravljanja Životnom Sredinom [Spatial Planning in the System of Environmental Protection]*; IAUS: Belgrade, Serbia, 2008; Available online: http://raumplan.iaus.ac.rs/handle/123456789/544 (accessed on 1 June 2022).
14. Crnčević, T.; Marić, I.; Josimović, B. Strategic environmental assessment and climate change in the Republic of Serbia: Support to development and adjustment process. *Spatium* **2011**, *26*, 14–19. [CrossRef]
15. Linkov, I.; Satterstrom, F.K.; Kiker, G.; Batchelor, C.; Bridges, T.; Ferguson, E. From comparative risk assessment to multi-criteria decision analysis and adaptive management: Recent developments and applications. *Environ. Int.* **2006**, *32*, 1072–1093. [CrossRef]
16. Nilsson, M.; Dalkmann, H. Decision making and strategic environmental assessment. *J. Environ. Assess. Policy Manag.* **2001**, *3*, 305–327. [CrossRef]
17. Proctor, W.; Drechsler, M. Deliberative multicriteria evaluation. *EPC* **2006**, *24*, 169–190. [CrossRef]
18. Josimović, B.; Marić, I.; Milijić, S. Multi-criteria evaluation in strategic environmental assessment for waste management plan, a case study: The city of Belgrade. *Waste Manag.* **2015**, *36*, 331–342. [CrossRef]
19. Rosen, L.; Norrman, J.; Norberg, T.; Volchko, Y.; Soderqvist, T.; Back, P.-E.; Norin, M.; Brinkhoff, P.; Bergknut, M.; Doberl, G. SCORE: Multi-criteria analysis (MCA) for sustainability appraisal of remedial alternatives. In Proceedings of the Second International Symposium on Bioremediation and Sustainable Environmental Technologies, Jacksonville, FL, USA, 10–13 June 2013; Available online: https://publications.lib.chalmers.se/records/fulltext/183067/local_183067.pdf (accessed on 1 June 2022).
20. Sparrevik, M.; Barton, D.N.; Bates, M.E.; Linkov, I. Use of stochastic multi-criteria decision analysis to support sustainable management of contaminated sediments. *Environ. Sci. Technol.* **2011**, *46*, 1326–1334. [CrossRef]
21. Shammi, M.; Rahman, M.M.; Ali, M.L.; Khan, A.S.M.; Siddique, M.A.B.; Ashadudzaman, M.; Bodrud-Doza, M.; Alam, G.M.M.; Tareq, S.M. Application of short and rapid strategic environmental assessment (SEA) for biomedical waste management in Bangladesh. *Case Stud. Chem. Environ. Eng. (CSCEE)* **2022**, *5*, 100177. [CrossRef]
22. Jay, S. Strategic environmental assessment for energy production. *Energy Policy* **2010**, *38*, 3489–3497. [CrossRef]
23. Josimović, B.; Crnčević, T. Impact evaluation within strategic environmental assessment: The case study of the waste management regional plan for Kolubara region in Serbia. *Environ. Eng. Manag. J. (EEMJ)* **2009**, *8*, 457–462. [CrossRef]
24. Calvo, F.; Moreno, B.; Zamorano, M.; Szanto, M. Environmental diagnosis methodology for municipal waste landfills. *Waste Manag.* **2005**, *25*, 768–779. [CrossRef]

25. Tchobanoglous, G.; Theisen, H.; Vigil, S.A. *Integrated Solid Waste Management: Engineering Principles and Management Issues*; McGraw-Hill: New York, NY, USA, 1993.
26. Josimović, B.; Marić, I.; Manić, B. Metodološki pristup u određivanju lokacije deponije komunalnog čvrstog otpada-studija slučaja-regionalna deponija za Kolubarski region [Methodological approach to the determination of landfill location for municipal solid waste: Case study: Regional landfill in Kolubara region]. *Arhit. I Urban.* **2011**, *32*, 55–64. [CrossRef]
27. Liou, M.-L.; Yeh, S.-C.; Yu, Y.-H. Reconstruction and systemization of the methodologies for strategic environmental assessment in Taiwan. *Environ. Impact Assess. Rev.* **2006**, *26*, 170–184. [CrossRef]
28. Krunić, N.; Josimović, B.; Gajić, A.; Nenković-Riznić, M. Territorial analysis as support to the strategic environmental assessment process for agro-waste management planning. *Spatium* **2019**, *42*, 16–22. [CrossRef]
29. Sheate, W.; Richardson, J.; Aschemann, R.; Palerm, J.; Steen, U. *SEA and Integration of the Environment into Strategic Decision-Making. (Main Report)*; Imperial College Consultants Ltd: London, UK, 2001; Volume 1, Available online: https://ec.europa.eu/environment/archives/eia/sea-studies-and-reports/pdf/sea_integration_main.pdf (accessed on 1 June 2022).
30. Marsden, S. Strategic environmental assessment: An international overview. In *Strategic Environmental Assessment in Australasia*; Marsden, S., Dovers, S., Eds.; The Federation Press: Sydney, Australia, 2002; pp. 1–23.
31. Josimović, B.; Cvjetić, A.; Furundžić, D. Strategic environmental assessment and the precautionary principle in the spatial planning of wind farms–European experience in Serbia. *Renew. Sustain. Energy Rev.* **2021**, *136*, 110459. [CrossRef]
32. Josimović, B.; Cvjetić, A.; Manić, B. Strategic environmental assessment in the application of preventive protection for wind farm noise—case study: Maestrale ring wind farm. *Energies* **2021**, *14*, 6174. [CrossRef]
33. Josimović, B.; Krunić, N.; Gajić, A.; Manić, B. Multi-criteria Evaluation in strategic environmental assessment in the creation of a sustainable agricultural waste management plan for wineries: Case study: Oplenac Vineyard. *J. Agric. Environ. Ethics* **2021**, *34*, 4. [CrossRef]
34. Josimović, B.; Krunić, N. Implementation of GIS in selection of locations for regional landfill in the Kolubara region. *Spatium* **2008**, *17–18*, 72–77. [CrossRef]
35. Pravilnik o Nacionalnoj listi indikatora zaštite životne sredine [Rulebook on the National List of Environmental Indicators]. *Off. Gaz. Repub. Serb.* **2011**, *37*, 99–228.
36. Chatterjee, N.C.; Bose, G.K. Selection of vendors for wind farm under fuzzy MCDM environment. *Int. J. Ind. Eng. Comput.* **2013**, *4*, 535–546. [CrossRef]
37. Rzeszutek, M.; Szulecka, A.; Oleniacz, R.; Bogacki, M. Assessment of the AERMOD dispersion model over complex terrain with different types of meteorological data: Tracy Power Plant experiment. In Proceedings of the E3S Web Conf. 22. International Conference on Advances in Energy Systems and Environmental Engineering (ASEE17), Wrocław, Poland, 2–5 July 2017. [CrossRef]
38. Ismail, W.K.W.; Abdullah, L. A new Environmental Performance Index using analytic hierarchy process: A case of ASEAN countries. *Environ. Skept. Crit.* **2012**, *1*, 39–47. Available online: http://www.iaees.org/publications/journals/environsc/articles/2012-1(3)/a-new-environmental-performance-index.pdf (accessed on 1 June 2022).
39. Palafox-Alcantar, P.G.; Hunt, D.V.L.; Rogers, C.D.F. A Hybrid methodology to study stakeholder cooperation in circular economy waste management of cities. *Energies* **2020**, *13*, 1845. [CrossRef]
40. Josimović, B. *Spatial Aspects of the Impact of Wind Farms on the Environment*; IAUS: Belgrade, Serbia, 2020; pp. 1–184. Available online: http://raumplan.iaus.ac.rs/handle/123456789/545 (accessed on 1 June 2022).
41. Kochańska, E.; Łukasik, R.M.; Dzikuć, M. New circular challenges in the development of take-away food packaging in the COVID-19 period. *Energies* **2021**, *14*, 4705. [CrossRef]

Article

Determining the Effectiveness of Street Cleaning with the Use of Decision Analysis and Research on the Reduction in Chloride in Waste

Anna Gronba-Chyła [1,*], Agnieszka Generowicz [2], Paweł Kwaśnicki [1,3], Dawid Cycoń [3], Justyna Kwaśny [2], Katarzyna Grąz [1], Krzysztof Gaska [4] and Józef Ciuła [5]

1. Faculty of Natural Sciences and Health, John Paul II Catholic University of Lublin, ul. Konstantynów 1 H, 20-708 Lublin, Poland; pawel.kwasnicki@kul.pl (P.K.); katarzyna.graz@kul.pl (K.G.)
2. Department of Environmental Technologies, Cracow University of Technology, ul. Warszawska 24, 31-155 Cracow, Poland; agenerowicz@pk.edu.pl (A.G.); kwasny.justyna@gmail.com (J.K.)
3. Research & Development Centre for Photovoltaics, ML System S.A., Zaczernie 190G, 36-062 Zaczernie, Poland; pawel.kwasnicki@mlsystem.pl
4. Department of Water and Wastewater Engineering, Silesian University of Technology, ul. Konarskiego 18, 44-100 Gliwice, Poland; krzysztof.gaska@polsl.pl
5. Institute of Engineering, State University of Applied Sciences in Nowy Sącz, ul. Zamenhofa 1A, 33-300 Nowy Sącz, Poland; jciula@pwsz-ns.edu.pl
* Correspondence: amgronba@kul.pl

Citation: Gronba-Chyła, A.; Generowicz, A.; Kwaśnicki, P.; Cycoń, D.; Kwaśny, J.; Grąz, K.; Gaska, K.; Ciuła, J. Determining the Effectiveness of Street Cleaning with the Use of Decision Analysis and Research on the Reduction in Chloride in Waste. *Energies* **2022**, *15*, 3538. https://doi.org/10.3390/en15103538

Academic Editor: Andrea De Pascale

Received: 24 March 2022
Accepted: 10 May 2022
Published: 12 May 2022

Publisher's Note: MDPI stays neutral with regard to jurisdictional claims in published maps and institutional affiliations.

Copyright: © 2022 by the authors. Licensee MDPI, Basel, Switzerland. This article is an open access article distributed under the terms and conditions of the Creative Commons Attribution (CC BY) license (https://creativecommons.org/licenses/by/4.0/).

Abstract: Waste from street cleaning is usually of a fine fraction below 10 mm and varies greatly in both quantity and composition. It may be composed of chlorides, especially for that resulting during winter due to the use of street de-icing agents. Chlorides can cause the salinization of surface water and groundwater, and the salinization of soils, which in turn lead to the deterioration of water purity and a decrease in biodiversity of aquatic organisms, changes in microbiological structure, and increases in toxicity of metals. Therefore, it is very important to determine the level of salinity in stored waste and its impact on the environment. The present study was conducted in a city of about 55,000 inhabitants. The highest chloride concentrations were observed after winter in waste from street and sidewalk cleaning around the sewer gullies, amounting to 1468 mg/dm^3. The lowest chloride concentration in this waste occurred in summer and amounted to 35 mg/dm^3. The multi-criteria analysis indicated that the most beneficial form of street cleaning and, thus, of reductions in chloride concentration in the waste from street cleaning, would be sweeping and daily washing. The objective of this research was to determine the amount of chlorides in sweepings on an annual basis in order to determine the potential risks associated with their impact on select aspects of the environment and to evaluate the frequency of necessary cleaning for city streets, considering the effects. The methodology used was a multi-criteria evaluation, which as a decision analysis, allowed us to determine the frequency of cleaning and washing of streets, in such a way that an ecological effect is achieved with simultaneous economic efficiency.

Keywords: street pollution; street cleaning; waste; chloride content; decision analysis; multi-criteria analysis

1. Introduction

According to the World Bank's report on the future of solid waste management, the world generates about 2.01 billion tons of solid waste annually and 33% (663 million tons) is not managed in an environmentally safe manner. According to research by the International Solid Waste Association (ISWA), 70% of municipal waste worldwide becomes landfill without treatment. By 2050, the amount of waste in the world will increase by 70%, reaching 3.40 billion tons of solid waste generated annually.

Waste management should be carried out according to the following hierarchy: prevention at the source, preparation for reuse, recycling and reuse of usable materials, other recovery methods including incineration, and final disposal of treatment residues. There is a wide variety of waste streams in large metropolitan areas, and this diversity creates the need for mass reduction and the need to dispose of all streams, regardless of their nature. Building and managing comprehensive systems would allow for a solution for waste management to be found, and by managing them, emissions to the environment can be reduced while recovering the raw material fraction and following the "polluter pays" principle. The basis for the creation of an objective function in the research on the optimization of regional waste management is the economic efficiency index for a complex system, which is extremely important because of the necessity to establish fees for waste management in accordance with the "polluter pays" principle. It is also important to reconcile the ecological and economic balance in such a way that the activities carried out balance both issues.

2. Chlorides and Their Impact on the Environment

One of the municipal waste streams is street sweepings generated during street cleaning and washing processes. What is interesting is both their quantity as well as their composition and variability in characteristics over the course of the year, which have a significant impact on the quality of the natural environment as they may affect the quality of the soil, water, and municipal infrastructure (roads, sidewalks, sewers, cars, etc.). According to the European Waste Catalogue [1], street cleaning waste, classified as "20 03 03–street-cleaning residues", is treated as one of the municipal waste streams that should be collected and disposed of in a comprehensive system. This waste may constitute 10–15% of the mass of municipal waste. However, the amount of suspended solids washed out during street cleaning and washing processes, which also has a significant impact on the quantity and quality of waste, has not been taken into account in previous studies. According to research [2,3], the amount of suspended solids represents 7–75% of the solid waste in relation to the amount of sweepings collected on particular days, only during sweeping. In general, the most commonly studied phenomenon reported in the literature is the impact of street cleaning on air quality and secondary emissions. They do not present conclusive results. Some studies conducted in this regard have registered an increase in PM10 (Particulate matter) levels [4,5] and an increase in the proportion of mineral components in particulate matter especially in the PM10 fraction [6]. On the other hand, it has been proven that only street sweeping may have periodical adverse effects on the removal of pollutants from the air; Vaze and Chiew [7] found that there were more fine dust particles in the air after street sweeping compared with before sweeping. In addition to studying street cleaning by sweeping alone, street washing and its effects on ambient air quality have also been studied. Some studies have shown that the effectiveness of street washing (without sweeping) is more related to the wetting effect than to the effective removal of particulate matter. On paved roads, the effect of street washing on air quality has been studied in Germany and Scandinavian countries [4,5,7]. The results showed that the effectiveness depends strongly on the local situation (location, meteorology, and road quality). Results presenting the effectiveness of street sweeping and washing are also presented in Reference [8], where the concentration of suspended particulate matter was controlled. References [9,10] describe the evaluation of the effectiveness of mechanical sweeping and street washing with water to reduce PM10 concentrations in ambient air. A significant number of publications describe the negative effects of the composition of sweepings deposited on city streets and in urban areas on the environment of these areas. Reference [11] analyzed dust collected from streets and soil from cities with high, medium, and low population densities and in a non-urbanized area. That study concluded that high population density increases the salinity of sweepings and soil but has no effect on the concentration of metals in soil. Reference [12] presented the results of a study on the identification of contaminants found in street dust from London (UK), New York (USA), Halifax (Canada), Christchurch (New Zealand), and Kingston (Jamaica). The pollutants identified were divided into two groups: of soil origin and from other sources, including tire wear, car emissions, and salt use. That study showed that the concentrations of most elements increase with a decrease in dust particle size. The salinity of the sweepings consequently increases the salinity of wastewater, either from street cleaning or runoff during precipitation, and the increased chloride content interferes with the dephosphatation and deflocculation process [13]. Due to a lack of management technology, this waste is deposited in landfills [14–17], consequently causing salinization of the landfill leachate [18,19]. De-icing agents are common road-safety maintenance materials during winter. According to the Polish law [20], the following can be used to maintain roads in winter: NaCl, $CaCl_2$, and $MgCl_2$. In addition, sand is used for road maintenance during winter in Poland to improve grip. On average, more than 500 thousand tons of sodium chloride is applied to roads in Poland [21]. For comparison, a country such as Sweden, where snowfalls are significantly higher, does not exceed the consumption of road salt above the level of 300 thousand tons per year. In the United States, on the other hand, 20 million tons of road salt is used annually. As the snow and ice melt, the salt is washed away and, together with precipitation or street cleaning water, it ends up in the ground, groundwater and surface water, polluting them. The greatest chloride leaching occurs during storms and downpours. It affects entire ecosystems. A heavy influx of chloride ions disrupts the ability of freshwater organisms to regulate fluid flow [22]. Changes in the salinity of a pond or lake can also affect the way water mixes with the change of seasons, leading to the formation of salt pockets near the bottom and biological dead zones. Increased salinity in water bodies can lead to decreased biodiversity of aquatic organisms, changes in microbial structure, and increased metal toxicity [23]. Increased chloride concentrations also contribute to groundwater salinity [24]. According to the Polish law, concentrations in groundwater for quality class I waters must not exceed 60 mg/dm^3, those for class II must not exceed 150 mg/dm^3, and those for class III must not exceed 250 mg/dm^3 [25]. For surface water, the chlorides concentration for class I must not exceed 5 mg/dm^3, while those for class II must not exceed 8.2 mg/dm^3, and there are no standards for other classes [26]. When road salt runs off the road, it can destroy soil, trees, and vegetation or limit their growth [27] up to 100 m from where the salt is spread. In addition, it corrodes sewage infrastructure and erodes road surfaces [28]. Roadside roads can also turn into artificial licks, attractive to animals such as deer and elk, increasing the risk of accidents [29–31].

In this article, an assessment of the chloride removal strategy from the environment in washing and cleaning processes is made based on a decision analysis (multi-criteria). Multi-criteria decision models have been used since the 1980s as an environmental impact assessment tool in environmental engineering [32–46]. The objective of the research presented was to determine the amount of chlorides in the sweepings on an annual basis in order to determine the potential risks associated with their impact on select aspects of the environment and to evaluate the frequency of cleaning urban streets, taking into account environmental and economic effects.

The methodology used is a multi-criteria evaluation, which as a decision analysis, allows use to determine the frequency of cleaning and washing of streets, in such a way as that an ecological effect is achieved with simultaneous economic efficiency.

3. Proposed Research Methodology

The scientific objective of the proposed methodology was to use a multi-criteria assessment to assess the degree of street pollution and to determine the optimal solution from the perspective of economic and ecological aspects.

The utilitarian/practical goal was to create a decision-making tool for waste management processes, taking into account environmental issues.

In the proposed research approach, a research methodology that allows us to determine the amount of chlorides in street cleaning waste and to determine the potential environmental risks associated with their presence is presented. In terms of the works undertaken, we propose the following:

- a selection of test times and places, and a selection of sampling locations;

- laboratory testing of the samples, processing of the test results, and conclusions from the analytical tests; and
- a decision analysis and a selection of the system of treatments and removal of chlorides from the environment according to the following scheme: development of treatment options taking into account environmental quality studies with costs, proposal of conditions and limitations to the decision analysis, and identification of the most beneficial scenario taking into account environmental and economic factors.

The necessary condition for finding the solution is a set of well-designed criteria that can be used for the evaluation of different options. The criteria taken into account should represent diverse goals that sometimes are even contradictory, e.g., a solution that is the cheapest and, at the same time, the most reliable. Thus, the options analyzed should be defined in detail, and the final selection is always a compromise based on the relative weights assigned to individual criteria. For the multi-criteria analysis, the compromise programming method was used, using the concept of organizing individual variants of technology modernization according to their distance from an established ideal point $X'(x_1'$, $x_2', \ldots, x_m')$, all x_m' coordinates of which are equal to the maximum value of the adopted normalization scale. The utility of s_n strategy with regard to all criteria can be expressed as follows:

$$U(s_n) = \sum_{m=1}^{M} w_m \cdot (r'_{nm} - x*_m); m = 1, \ldots, M \tag{1}$$

where

$U(s_n)$—s_n strategy utility function;
n—number of strategies;
m—number of criteria;
W_m—the weight of each criterion, assumed by the decision-maker;
r'_{nm}—standardized evaluation criterion; and
$x*_m$—mth nadir coordinate, which is the most unfavorable strategy.

The search for the most favorable strategy is carried out according to the following rule:

$$s_j \Leftrightarrow U(s_j) = \max U(s_n); n = 1, \ldots, N \tag{2}$$

where

s_j—the most advantageous technological variant sought.

The applied method leads to a complete ordering of the elements of the decision area and finding the most environmentally beneficial solution.

4. Description and Results of Analytical Research

The research was conducted in a medium-sized city with a population of about 55,000 inhabitants. The area from which street cleaning waste was removed is about 140 km^2, while the area of the sidewalks is about 450 thousand m^2. The amount of sweepings collected annually ranges from 300 to 1200 tons per year, depending on the severity of winter in a given year. The ratio in which sand is mixed with NaCl during winter is 50:50; at temperatures below 20 °C, $CaCl_2$ is also used at a ratio of 50:50 with sand. Sidewalks are only gritted with sand, without salt. After collection, this waste is deposited in a landfill for non-inert and hazardous waste. The fee for depositing this waste is EUR 65 per ton.

Samples for the research were taken during two periods of increased street and sidewalk cleaning, that is, the end of summer—August/September—and the end of winter—March. Sampling for laboratory tests (solid waste and street washing wastewater samples) were taken directly from the waste container of the street and sidewalk cleaning truck according to the standards [47,48].

The maximum amount of street cleaning waste collected was 270 kg/km, while the minimum was 200 kg/km. The content of organic parts was 9% and the content of mineral parts was 91%, which are shown in Figure 1.

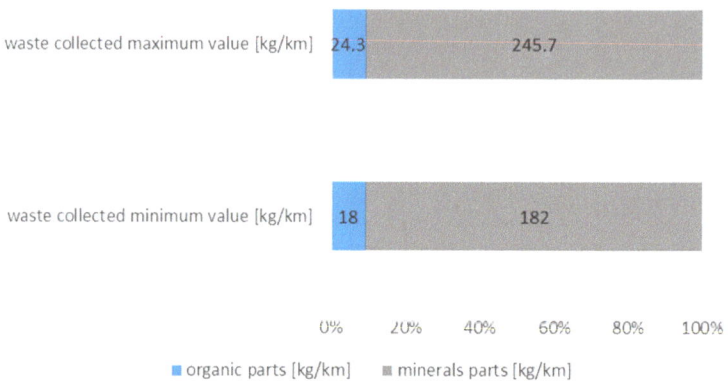

Figure 1. Maximum and minimum amounts of street cleaning waste collected, divided into organic and mineral components.

The maximum collected suspended solids from street cleaning was 26 mg/dm^3, while the minimum was 12 mg/dm^3. The organic content of the suspended solids was 17% and the mineral content was 83%, which are shown in Figure 2.

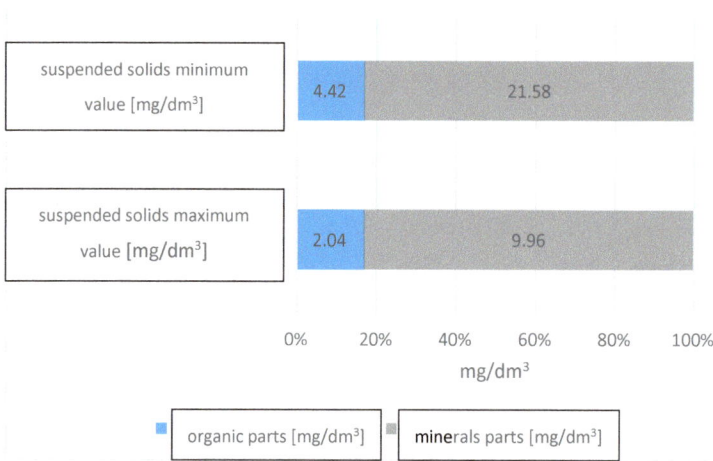

Figure 2. Maximum and minimum amounts of collected suspended solids, divided into organic and mineral components.

Due to the possibility of salt accumulation in different areas of roads, we proposed to take samples not only from streets but also from wastewater and gullies. The samples labeled Street 1 and Sidewalk are solid samples. Samples for the research were taken from this waste by dissolving about 200 g of waste in distilled water and then filtering it, and a representative sample was obtained for testing. On the other hand, samples from Street 2 from around the gullies and samples from Street 3 (streets washing) were taken in semi-liquid form and filtered, and a representative sample was obtained for testing. For each sampling location—Street 1, Street 2, Street 3, and Sidewalk—30 representative samples were obtained and analyzed for chloride content. The chloride contents of wastewater and street sweepings were determined using the standard: PN-ISO 9297:1994 [49]. The results for the chloride content in 30 samples for each site are presented in Table 1 and Figure 3.

Table 1. Chloride content in individual street and sidewalk cleaning samples after summer and winter.

Site	Chloride Content in Street Cleaning Wastewater (mg/dm^3) after Summer		Chloride Content in Street Cleaning Wastewater (mg/dm^3) after Winter	
	Minimum Value	Maximum Value	Minimum Value	Maximum Value
Street 1	29.3	41.4	321.6	399.8
Street 2 (gullies)	45.6	71.9	1201.4	1732.8
Street 3 (washing)	38.3	49.3	436.3	468.5
Sidewalk	25.8	46.2	45.9	59.2

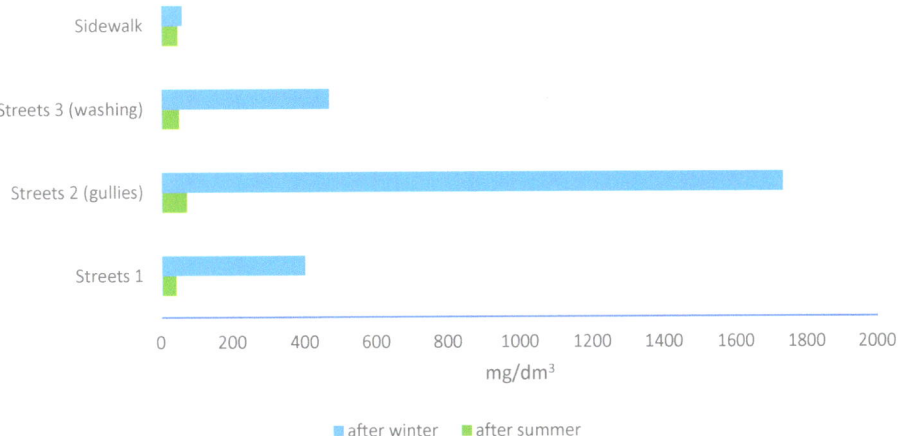

Figure 3. Comparison of chloride content (mg/dm^3) on select streets in cleaning sweepings collected after summer and winter (average values).

The study showed significant differences in the amount of chloride in the "after winter" and "after summer" periods. This is understandable given the nature of the periods in which the study was conducted. In the samples collected for testing in the "after summer" period, the highest average chloride concentration of about 60 mg/dm^3 was obtained in the Street 2 samples collected from the vicinity of sewer gullies. This was followed by an average concentration of about 44 mg/dm^3 obtained in the Street 3 samples collected from street washing. Similar average concentrations were obtained in the Street 1 samples, approximately 35 mg/dm^3, and in the Sidewalk samples, approximately 36 mg/dm^3.

During the post-winter season, the average chloride concentrations peaked at approximately 1468 mg/dm^3 in the Street 2 samples collected from around sewer gullies. In the Street 3 samples (street washing), the average chloride concentration was approximately 422 mg/dm^3. In the Street 1 samples, the average chloride concentration was approximately 362 mg/dm^3. The lowest chloride concentration from this period was obtained in the Sidewalk samples, about 53 mg/dm^3 on average.

From the results obtained in these two seasons, a correlation in the concentration levels is observed at the respective sampling sites. The highest chloride concentrations are found in samples collected near gullies and the lowest chloride concentrations are found in samples collected from sidewalks.

The low chloride concentrations on the sidewalks are due to a ban on salt gritting the sidewalks. The high concentration of chlorides at sewer gullies is due to runoff into the sewer system and the accumulation of street runoff along with sweepings near the gullies. This high average concentration of chlorides present in street cleaning residues is due to the long, snowy, and cold winter, during which roads were abundantly gritted with sand and salt.

5. Results of Decision Analysis Using Multi-Criteria Analysis

The decision problem was formulated when the evaluation criteria were established and their values were expressed in the form of a finite set of numbers (measurable values), which are the result of evaluating the different variants of the proposed chloride removal system from an urbanized area, against the selected criteria. The presented strategies described by measurable criteria constitute a decision matrix (Table 2). The values of emissions to the environment as a result of the maintenance process in the study area are presented in the columns of Table 2. Among the studied strategies, the following were proposed:

Table 2. Decision matrix for evaluating the adopted scenarios for the system of chloride removal from the environment.

Criteria	Unit	Sweeping	Sweeping + Washing 1 Day	Sweeping + Washing 2 Days	Sweeping + Washing 5 Days
		Scen1	Scen2	Scen3	Scen4
waste amount	kg/km	200	200	250	270
amount of wastewater from street cleaning	l/km	0	25	50	75
reduction in suspended solids on subsequent cleaning days		0	30	45	67
costs	euro/km	3	5	7.5	10
chloride content reduction		0	56	78	90

Scen1—proposing street cleaning only by sweeping followed by 1 week off;
Scen2—proposing street cleaning by sweeping and washing on 1 day followed by 1 week off;
Scen3—proposing street cleaning by sweeping and washing on the following 2 days, followed by 1 week off; and
Scen4—proposing street cleaning by sweeping and washing on the following 5 days, followed by 1 week off.

Among the criteria for evaluating the presented scenarios, the following were suggested:
1. Amount of waste collected cumulatively in consecutive days of cleaning measured in kg/technological km;
2. Amount of wastewater from street cleaning, cumulatively over successive days of street cleaning, measured in l/technological km;
3. Reduction in suspended solids calculated as the result of the difference between the first (on the first day) and the last street washing (on the last day);
4. Costs of each consecutive street cleaning activity, including equipment maintenance costs, water and fuel consumption, employee costs, and environmental emissions fees, that were estimated on the basis of actual measurements; and
5. Reduction in chloride content calculated as the difference in chloride content between the first and last days of cleaning and washing.

The matrix defined in this way became the formulated decision problem to be solved, in which the compromise programming method expressed by formulas (1) and (2) was used. The results and final ordering of the individual maintenance strategies, with a particular emphasis on chloride removal, are presented in Table 3, ranking them from most to least favorable. The ranking depends additionally on the adopted weights of each group of criteria or individual criteria. In Table 3, the first column presents the weights of the criteria proposed by the authors of this research. In most cases, these weights were given to individual criteria described in Table 3. Thus, in the first row of Table 3, all criteria were given a weight of 1, while in the second row, the first criterion was given a weight of 2, while the remaining criteria were given weights of 1. In the last two rows of Table 3, only the criterion assessing the costs of individual solutions received a lower weight than the remaining ones.

Table 3. Multi-criteria analysis results ranking maintenance and chloride removal scenarios.

Validity of the Criteria	Ranking of Scenarios		
	$\alpha = 1$	$\alpha = 2$	$\alpha = \infty$
1:1:1:1:1	Scen2→Scen3→Scen4→Scen1	Scen2→Scen3→Scen4→Scen1	Scen2↔Scen3→Scen1↔Scen4
2:1:1:1:1	Scen3→Scen2→Scen4→Scen1	Scen2→Scen3→Scen4→Scen1	Scen2↔Scen3→Scen1↔Scen4
10:1:1:1:1	Scen4→Scen3→Scen2→Scen1	Scen3→Scen4→Scen2→Scen1	Scen2↔Scen3→Scen1↔Scen4
1:2:1:1:1	Scen2→Scen1→Scen3→Scen4	Scen2→Scen1→Scen3→Scen4	Scen2→Scen3→Scen1
1:10:1:1:1	Scen1→Scen2→Scen3→Scen4	Scen1→Scen2→Scen3→Scen4	Scen2→Scen3→Scen1
1:1:2:1:1	Scen4→Scen3→Scen2→Scen1	Scen3→Scen2→Scen4→Scen1	Scen2↔Scen3→Scen4
1:1:10:1:1	Scen4→Scen3→Scen2→Scen1	Scen4→Scen3→Scen2→Scen1	Scen2↔Scen3→Scen4
1:1:1:2:1	Scen2→Scen1→Scen3→Scen4	Scen2→Scen1→Scen3→Scen4	Scen2→Scen3→Scen1
1:1:1:10:1	Scen1→Scen2→Scen3→Scen4	Scen1→Scen2→Scen3→Scen4	Scen2→Scen3→Scen1
1:1:1:1:2	Scen3→Scen4→Scen2→Scen1	Scen3→Scen2→Scen4→Scen1	Scen2↔Scen3→Scen4
1:1:1:1:10	Scen4→Scen3→Scen2→Scen1	Scen4→Scen3→Scen2→Scen1	Scen2↔Scen3→Scen4
2:2:2:1:2	Scen4→Scen3→Scen2→Scen1	Scen2→Scen3→Scen4→Scen1	Scen2↔Scen3
10:10:10:1:10	Scen4→Scen3→Scen2→Scen1	Scen3→Scen2→Scen4→Scen1	Scen2↔Scen3

Summarizing the results of the analysis, the following should be noted:

- It is possible to select a cleaning scenario for a selected area using a decision analysis, proposing environmental and economic criteria for evaluation.
- According to the results of the calculations presented in Table 3, it can be seen that the most frequently selected scenario is Scen2 with sweeping and one-day street washing.
- Scen4, sweeping and washing the street in a 5-day system, is selected as the most advantageous eight times (including two times while the weight of the cost criterion was reduced in relation to environmental criteria), which allows us to conclude that costs should be taken into account and calculated in the selection and analysis of cleaning strategies each time.
- Scen1 (sweeping only) and Scen4 (where the cleaning and washing process is the longest 5-day process) are selected as the least favorable in a significant number of cases; therefore, it can be said that, by balancing the economic and ecological effects, a scenario that allows for an observance of the principles of sustainable development is selected.
- This method gives the possibility of additional weighting of the criteria by using the α exponent in formula (2). This exponent allows for additional weighting of each deviation from the ideal point in proportion to their value. The larger the value of α is, the greater the importance of large deviations of the strategy from the ideal point. For $\alpha = \infty$, scenario 2 is always selected as the most favorable.

6. Conclusions

1. The results of this research carried out in a medium city show how much chloride is present in sweepings. The diversification of sampling locations allowed us to determine where pollutants accumulate. These differences are very significant and amount to almost 1050 mg/dm^3 in the period after winter between the place of sampling near the sewage gully and the roadway. Such a high concentration of chlorides in the sweepings deposited in the landfill in large quantities can cause an increase in the salinity of the landfill leachate, which in turn affects the prolonged decomposition of matter, the formation of biogas, and the possibility of salinization of groundwater and impede biological processes in sewage-treatment plants.
2. The selection of a cleaning strategy with a particular emphasis on chloride reduction in this region is a difficult decision task that must take into account various, often conflicting goals and objectives, and socioeconomic interests. The measuring criteria defined allowed us to establish a quantitative and objectified evaluation of the performance of such a system. The proposed methodology provides the possibility of a quantitative; multifaceted; and at the same time, objectified evaluation of scenario solutions, replacing intuitive evaluations or those requiring expert opinions used so far. In the proposed example, Scen2, which includes sweeping and one-time street cleaning in a 1-day system, was selected as the most beneficial.
3. The proposed methodology allows us to evaluate the system on an ongoing basis in accordance with the requirements of environmental management even if the objective or conditions in the region change.

Author Contributions: Conceptualization, A.G.-C., A.G., J.K., K.G. (Krzysztof Gaska), P.K., D.C., K.G. (Katarzyna Grąz) and J.C.; methodology, A.G.-C., A.G., J.K., K.G. (Krzysztof Gaska), P.K., D.C., K.G. (Katarzyna Grąz) and J.C.; software, A.G.-C. and A.G., validation, A.G.-C., A.G., K.G. (Krzysztof Gaska) and J.C.; formal analysis, A.G.-C., A.G., J.K. and K.G. (Katarzyna Grąz); investigation, A.G.-C., A.G., J.K., K.G. (Krzysztof Gaska), P.K., D.C., K.G. (Katarzyna Grąz) and J.C.; resources, A.G.-C., A.G., P.K., D.C., K.G. (Katarzyna Grąz) and J.C; data curation, A.G.-C., A.G., J.K., K.G. (Krzysztof Gaska), K.G. (Katarzyna Grąz) and J.C; writing—original draft preparation, A.G.-C. and A.G.; writing—review and editing, A.G.-C. and A.G.; visualization, A.G.-C. and A.G.; supervision, A.G.-C. and A.G.; project administration, A.G.-C. and A.G.; funding acquisition, A.G.-C. and A.G. All authors have read and agreed to the published version of the manuscript.

Funding: This work was supported by The National Centre for Research and Development grant no: POIR.01.02.00-00-0265/17-00 "Application of low-dimensional structures for broadening the absorption spectrum and enhancing the efficiency of silicon cells in IBC or BIFACIAL architecture" Ml System S.A., 36-062 Zaczernie 190 G.

Institutional Review Board Statement: Not applicable.

Informed Consent Statement: Not applicable.

Data Availability Statement: https://repozytorium.kul.pl/handle/20.500.12153/25, accessed on 27 February 2022.

Conflicts of Interest: The authors declare no conflict of interest.

References

1. COMMISSION DECISION of 18 December 2014 Amending Decision 2000532EC on the List of Waste Pursuant to Directive 200898EC of the European Parliament and of the Council.Pdf. Available online: https://eur-lex.europa.eu/legal-content/EN/TXT/?uri=CELEX:32014D0955 (accessed on 27 February 2022).
2. Alwaeli, M. End-of-Life Vehicles Recovery and Recycling and the Route to Comply with EU Directive Targets. *Environ. Prot. Eng.* **2016**, *42*, 191–202. [CrossRef]
3. Generowicz, A.; Wassilkowska, A.; Kryłów, M. Qualitative composition of waste from street cleaning on the example of research carried out in Krakow. *Przemysl. Chem.* **2020**, *99*, 1312–1314. [CrossRef]
4. Gronba-Chyła, A.; Generowicz, A.; Kramek, A. Using Selected Types of Waste to Produce New Light Ceramic Material. *Pol. J. Environ. Stud.* **2021**, *30*, 2073–2083. [CrossRef]

5. Gronba-Chyła, A.; Generowicz, A. Municipal waste fraction below 10 mm and possibility of its use in ceramic building materials. *Przem. Chem.* **2020**, *99*, 1318–1321. [CrossRef]
6. Norman, M.; Johansson, C. Studies of Some Measures to Reduce Road Dust Emissions from Paved Roads in Scandinavia. *Atmos. Environ.* **2006**, *40*, 6154–6164. [CrossRef]
7. AIRUSE. *AIRUSE LIFE 11 ENV/ES/584 the Scientific Basis of Street Cleaning Activities as Road Dust Mitigation Measure*; AIRUSE: Leiden, The Netherlands, 2013.
8. Vaze, J.; Chiew, F. Experimental Study of Pollutant Accumulation on an Urban Road Surface. *Urban Water* **2002**, *4*, 379–389. [CrossRef]
9. Aldrin, M.; Hobæk Haff, I.; Rosland, P. The Effect of Salting with Magnesium Chloride on the Concentration of Particular Matter in a Road Tunnel. *Atmos. Environ.* **2008**, *42*, 1762–1776. [CrossRef]
10. Chang, Y.-M.; Chou, C.-M.; Su, K.-T.; Tseng, C.-H. Effectiveness of Street Sweeping and Washing for Controlling Ambient TSP. *Atmos. Environ.* **2005**, *39*, 1891–1902. [CrossRef]
11. Amato, F.; Querol, X.; Alastuey, A.; Pandolfi, M.; Moreno, T.; Gracia, J.; Rodriguez, P. Evaluating Urban PM10 Pollution Benefit Induced by Street Cleaning Activities. *Atmos. Environ.* **2009**, *43*, 4472–4480. [CrossRef]
12. Amato, F.; Querol, X.; Johansson, C.; Nagl, C.; Alastuey, A. A Review on the Effectiveness of Street Sweeping, Washing and Dust Suppressants as Urban PM Control Methods. *Sci. Total Environ.* **2010**, *408*, 3070–3084. [CrossRef]
13. Acosta, J.A.; Gabarrón, M.; Faz, A.; Martínez-Martínez, S.; Zornoza, R.; Arocena, J.M. Influence of Population Density on the Concentration and Speciation of Metals in the Soil and Street Dust from Urban Areas. *Chemosphere* **2015**, *134*, 328–337. [CrossRef] [PubMed]
14. Sobiecka, E. Thermal and Physicochemical Technologies Used in Hospital Incineration Fly Ash Utilization before Landfill in Poland. *J. Chem. Technol. Biotechnol.* **2016**, *91*, 2457–2461. [CrossRef]
15. Fergusson, J.E.; Ryan, D.E. The Elemental Composition of Street Dust from Large and Small Urban Areas Related to City Type, Source and Particle Size. *Sci. Total Environ.* **1984**, *34*, 101–116. [CrossRef]
16. Qiao, X.-X.; Dong, Y.; Lei, Y.-S.; Zhou, L.-X.; Liu, F.-W. Effect of Chloride Ions on Biological Oxidation of Pyrite and the Biomineralization Behavior in Wastewater System. *Chin. J. Ecol.* **2018**, *37*, 1685–1692. [CrossRef]
17. Gluba, T.; Olejnik, T.R.; Obraniak, A. Technology for producing washing agent in continuous process. *Przemysl. Chem.* **2015**, *94*, 1370–1374. [CrossRef]
18. Zhou, Y.; Huang, M.; Deng, Q.; Cai, T. Combination and Performance of Forward Osmosis and Membrane Distillation (FO-MD) for Treatment of High Salinity Landfill Leachate. *Desalination* **2017**, *420*, 99–105. [CrossRef]
19. Ciuła, J. Modelling the migration of anthropogenic pollution from active municipal landfill in groundwaters. *Archit. Civ. Eng. Environ.* **2021**, *14*, 81–90. [CrossRef]
20. Generalna Dyrekcja Dróg Krajowych i Autostrad—Generalna Dyrekcja Dróg Krajowych i Autostrad—Portal Gov.pl. Available online: https://www.gov.pl/web/gddkia (accessed on 19 March 2022).
21. Ordinance of the Minister of Environment of 27 October 2005 on the Types and Conditions of Using Means That Can Be Used on Public Roads, Streets and Squares. *J. Laws* **2005**, *230*, 1960.
22. Findlay, S.E.G.; Kelly, V.R. Emerging Indirect and Long-Term Road Salt Effects on Ecosystems. *Ann. N. Y. Acad. Sci.* **2011**, *1223*, 58–68. [CrossRef]
23. Mazur, N. Effects of Road Deicing Salt on the Natural Environment. *Eng. Environ. Prot.* **2015**, *18*, 449–458. (In Polish)
24. Bäckström, M.; Karlsson, S.; Bäckman, L.; Folkeson, L.; Lind, B. Mobilisation of Heavy Metals by Deicing Salts in a Roadside Environment. *Water Res.* **2004**, *38*, 720–732. [CrossRef] [PubMed]
25. Regulation of the Minister of Maritime Affairs and Inland Navigation of 11 October 2019 on the Criteria and Method of Assessing the Status of Groundwater Bod(Dz.U.2019.2148). Available online: https://isap.sejm.gov.pl/isap.nsf/DocDetails.xsp?id=WDU20190000868 (accessed on 27 February 2022).
26. Regulation of the Minister of Environment of 21 July 2016 on the Method of Classification of the State of Surface Water Bodies and Environmental Quality Standards for Priority Substances (Journal of Laws 2016, 1187). Available online: https://isap.sejm.gov.pl/isap.nsf/DocDetails.xsp?id=WDU20160001187 (accessed on 27 February 2022).
27. Safdar, H.; Amin, A.; Shafiq, Y.; Ali, A.; Yasin, R.; Sarwar, M.I. A Review: Impact of Salinity on Plant Growth. *Nat. Sci.* **2019**, *1*, 34–40. [CrossRef]
28. Kuosa, H.; Ferreira, R.M.; Holt, E.; Leivo, M.; Vesikari, E. Effect of Coupled Deterioration by Freeze–Thaw, Carbonation and Chlorides on Concrete Service Life. *Cem. Concr. Compos.* **2014**, *47*, 32–40. [CrossRef]
29. Czajka, A. Control of the State of the Soil Environment in the Vicinity of Communication Infrastructure. *Eliksir* **2017**, *2*, 8–12. (In Polish)
30. Sławiński, J.; Gołąbek, E.; Senderak, G. Influence of transport pollution on soil and cultivated vegetation of the wayside. *Inż. Ekol.* **2014**, *40*, 137–144. [CrossRef]
31. Casey, R.; Lev, S.; Snodgrass, J. Stormwater Ponds as a Source of Long-Term Surface and Ground Water Salinisation. *Urban Water J.* **2013**, *10*, 145–153. [CrossRef]
32. Satterstrom, F.; Kiker, G.; Batchelor, C.; Bridges, T.; Ferguson, E.; Linkov, I.; Satterstrom, F.K.; Kiker, G.; Batchelor, C.; Bridges, T.; et al. From Comparative Risk Assessment to Multi-Criteria Decision Analysis and Adaptive Management: Recent Developments and Applications. *Environ. Int.* **2007**, *32*, 1072–1093. [CrossRef]

33. Wilson, E.J.; McDougall, F.R.; Willmore, J. Euro-Trash. Searching Europe for a More Sustainable Approach to Waste Management. *Resour. Conserv. Recycl.* **2001**, *31*, 327–346. [CrossRef]
34. Morrissey, A.J.; Browne, J. Waste Management Models and Their Application to Sustainable Waste Management. *Waste Manag.* **2004**, *24*, 297–308. [CrossRef]
35. Li, Y.P.; Huang, G.H. An Inexact Two-Stage Mixed Integer Linear Programming Method for Solid Waste Management in the City of Regina. *J. Environ. Manag.* **2006**, *81*, 188–209. [CrossRef]
36. Aragonés-Beltrán, P.; Mendoza-Roca, J.; Bes-Piá, A.; García-Melón, M.; Parra-Ruiz, E. Application of Multicriteria Decision Analysis to Jar-Test Results for Chemicals Selection in the Physical-Chemical Treatment of Textile Wastewater. *J. Hazard. Mater.* **2009**, *164*, 288–295. [CrossRef] [PubMed]
37. Garfí, M.; Tondelli, S.; Bonoli, A. Multi-Criteria Decision Analysis for Waste Management in Saharawi Refugee Camps. *Waste Manag.* **2009**, *29*, 2729–2739. [CrossRef] [PubMed]
38. Shmelev, S.; Powell, J. Ecological–Economic Modelling for Strategic Regional Waste Management Systems. *Ecol. Econ.* **2006**, *59*, 115–130. [CrossRef]
39. Vego, G.; Kučar-Dragičević, S.; Koprivanac, N. Application of Multi-Criteria Decision-Making on Strategic Municipal Solid Waste Management in Dalmatia, Croatia. *Waste Manag.* **2008**, *28*, 2192–2201. [CrossRef]
40. Cossu, R. Waste Management, Energy Production, Healthcare: Amazing Similarities. *Waste Manag.* **2011**, *31*, 1671–1672. [CrossRef]
41. Generowicz, A.; Kowalski, Z.; Kulczycka, J.; Makara, A. Multi-Criteria Analysis for Optimization of Sodium Chromate Production from Chromic Waste. *Soil Air Water* **2011**, *39*, 688–696. [CrossRef]
42. Generowicz, A.; Gaska, K.; Hajduga, G. Multi-Criteria Analysis of the Waste Management System in a Metropolitan Area. In Proceedings of the 10th Conference on Interdisciplinary Problems in Environmental Protection and Engineering EKO-DOK 2018, Polanica-Zdroj, Poland, 16–18 April 2018; Volume 44, p. 43. [CrossRef]
43. Generowicz, A.; Kulczycka, J.; Kowalski, Z.; Banach, M. Assessment of Waste Management Technology Using BATNEEC Options, Technology Quality Method and Multi-Criteria Analysis. *J. Environ. Manag.* **2011**, *92*, 1314–1320. [CrossRef]
44. Gaska, K.; Generowicz, A. SMART Computational Solutions for the Optimization of Selected Technology Processes as an Innovation and Progress in Improving Energy Efficiency of Smart Cities—A Case Study. *Energies* **2020**, *13*, 3338. [CrossRef]
45. Ciuła, J.; Kozik, V.; Generowicz, A.; Gaska, K.; Bak, A.; Paździor, M.; Barbusiński, K. Emission and Neutralization of Methane from a Municipal Landfill-Parametric Analysis. *Energies* **2020**, *13*, 6254. [CrossRef]
46. Qureshi, M.E.; Harrison, S.R.; Wegener, M.K. Validation of Multicriteria Analysis Models. *Agric. Syst.* **1999**, *62*, 105–116. [CrossRef]
47. *Polish Standard PN-93/Z-15008/01*; Municipal Solid Waste. Testing of Fuel Properties. General Provisions. Polish Committee for Standardization: Płock, Poland, 1993.
48. *Polish Standard PN-93/C-87071*; Final (Laboratory) Sample. General Guidelines. Polish Committee for Standardization: Płock, Poland, 1993.
49. *Polish Standard PN-ISO 9297:1994*; Water Quality—Determination of chlorides—Method of Titration with Silver Nitrate in the Presence of chromate As Indicator (Mohr Method). Polish Committee for Standardization: Płock, Poland, 1994.

Article

Synoptic Risk Assessment of Groundwater Contamination from Landfills

Sonja Cerar *, Luka Serianz, Katja Koren, Joerg Prestor and Nina Mali

Geological Survey of Slovenia, Dimičeva Ulica 14, 1000 Ljubljana, Slovenia; luka.serianz@geo-zs.si (L.S.); katja.koren@geo-zs.si (K.K.); joerg.prestor@geo-zs.si (J.P.); nina.mali@geo-zs.si (N.M.)
* Correspondence: sonja.cerar@geo-zs.si

Abstract: Waste management in Europe has improved in recent years, reducing the amount of waste disposed at landfills. However, there are still many landfills in the countries. It is well known that landfills that do not have measures in place to control leachate entering groundwater can contaminate groundwater long after the landfill is closed. Collecting monitoring results from all landfills allows permitting and management agencies to improve action plans. This relies on a synoptic risk assessment that allows prioritization and milestones to be set for required actions. The developed method of synoptic risk assessment is based on a conceptual model of the landfill and the results of chemical groundwater monitoring tested at 69 landfills in Slovenia. The study confirms that most landfills have a direct or indirect impact on groundwater quality. All landfills were classified into three priority classes on the basis of the synoptic risk assessment. The results show that a total of 24 landfills have a clearly pronounced impact on groundwater. A total of 31 landfills have a less pronounced impact due to the favorable natural attenuation capacity of the soil or the technically appropriate design of the landfill itself. A total of 14 landfills have a less pronounced or negligible impact on groundwater.

Keywords: conceptual model; synoptic risk assessment; landfill; groundwater; chemical analysis

1. Introduction

Landfills represent a variety of potential local sources of environmental pollution. One of the main hazards related with landfilling is the generation of leachate [1–3]. After the leachate has permeated the waste deposits and reached the groundwater and surface water, it may cause contamination in the wider environment. Contamination in groundwater may remain for decades or even centuries. Because landfills have the potential to produce leachate for several hundred years, a proper operation of all landfill systems must be ensured during the period of landfill closure. Public concern and awareness of environmental protection has grown worldwide, which is also considered in the development of environmental legislation in various countries [4]. In Europe, the requirements of the Waste Framework Directive 2006/12/EC are designed to prevent or reduce, as much as possible, the negative effects of landfilling on the environment and any resulting risks to humans monitoring the potential impact of landfills on the environment, as well as on groundwater. Groundwater monitoring at landfills is a crucial measure in minimizing and controlling the risk of pollution and is one of the first instruments required and regulated by national or international legislation. The Landfill Directive (1991–1999/31/EC) established the rules for determining a significant change in groundwater quality due to the unfavorable environmental impact of a landfill, which should be determined on the basis of a representative hydrogeological conceptual model and the establishment of appropriate criteria before landfilling activities begin.

Several toxic and harmful contaminants can migrate directly from landfill leachate to groundwater [5–8], consisting of inorganic compounds, organic contaminants, and xenobiotics [9]. Leachate can also cause reductive redox potentials in groundwater and affect the alteration of individual substances, thus negatively affecting the chemical status of groundwater [10]. In recent years, numerous studies have been published on the determination of physicochemical parameters of leachate [11–13] and their impact on groundwater under waste fields [14–16]. In most cases, groundwater quality is assessed by determining chemical and/or microbiological parameters and comparing the data with existing threshold values [15,17,18]. Such an approach provides information only on specific pollutants and, therefore, provides little information on general water quality. A more comprehensive description of the groundwater quality in the vicinity of the landfill can be obtained by different summary indices based on measurements of individual parameters, such as leachate pollution index (LPI) [11,19], water quality index (WQI) [20] and a modified WQI called the landfill water pollution index (LWPI) [21,22], and Nemerow index (PI) [23]. In fact, the generation of quality indices allows a synthesis of the results of environmental quality assessment so that it can be understood by nonexperts such as the public/stakeholders involved in decision making. In the last decade, many researchers also focused on assessing the impact of landfilling on groundwater quality using the risk assessment method, which is an ever-evolving assessment tool. The literature review on landfill risk assessment conducted by Butt et al. [4] highlighted that there is no such holistic risk assessment methodology for landfill leachate that could help carry out the risk assessment process from the beginning (i.e., baseline study) to the end (i.e., hazard indices and risk quantification). One of the problems of landfill impact assessment is the determination of background conditions in the landfill area [8,16]. Several appropriate waste management strategies have been developed to support the assessment of background conditions at the local scale [24]. However, in many real conditions, the monitoring network was established after the start of landfill activities or without adequate knowledge of the in situ hydrogeological conditions. Previous studies [24–27] have indicated that natural backgrounds can only be determined from a representative geochemical set by sampling groundwater. High metal levels exceeding groundwater acceptable threshold values are broadly and pre-emptively correlated to local pollution, completely ignoring the geogenic natural background levels [8].

In Europe, as well as in Slovenia, the disposal of waste in landfills and the assessment of its impact on groundwater and surface water status are regulated by regulations derived from EU legislation (Waste Framework Directive 2006/12/EC). It was recognized most recently that most of the landfills have an impact on local groundwater quality status; therefore, it is also necessary to assess which of these landfills threaten the chemical status of groundwater bodies on a regional scale [28], which is very important for the state management of water resources. The impact of the landfill on groundwater status must be assessed on the basis of the chemical characteristics of groundwater, as well as on geogenic and other external parameters (reliability of the monitoring network, interaction between surface water and groundwater, and higher upgradient chemical concentrations in groundwater). A relevant assessment of the impact of the landfill on groundwater requires the analysis of the possible consequences caused by the impact of the landfill on the degradation of groundwater-dependent ecosystems and drinking water supply.

All these uncertainties have led us to develop a relevant approach to assess the impact of the landfill on groundwater status by considering conceptual models of the landfill site. The article presents a risk assessment of the pollution impact on groundwater from landfills based on the reviewed monitoring reports in Slovenia. All landfills were divided into classes according to the degree of risk of pollution, providing insight into the status of the landfills in relation to landfill management practices and providing the basis for alternative corrective actions.

The proposed method focuses on estimating the relationship between key chemical parameters, and it also considers conceptual models of the landfills. The study had the following objectives: (I) to develop an integrated method for classifying landfills on the basis of groundwater chemical data and conceptual models of landfills, (II) to identify the main groundwater pollutants in the vicinity of landfills according to type and status of the landfills, (III) to assess the impact of landfills on groundwater in the case of Slovenia, and (IV) to classify landfills in Slovenia according to developed integrated method.

2. Material and Methods

2.1. Study Area

Slovenia is a European country that is surrounded by the Alps, Mediterranean Sea, and Central Europe. With an area of 20,273 km^2, Slovenia is one of the smallest European countries. It has about two million inhabitants.

In 2019, about 8.4 million tons of waste were generated, of which about 1.1 tons or 509 kg per capita was municipal waste with an annual increasing trend (426 kg per capita in 2002). Approximately 775,000 tons or 73% of municipal waste was collected separately, and this proportion also increased from 8.6% in 2002 to 73% in 2019, while the proportion of landfilled municipal waste decreased (from 84% in 2002 to 6.3% in 2019) due to the increased waste recycling rate (from 63% in 2010 to 85% in 2019) [29].

Slovenia has some of the richest groundwater resources in Europe. Over 97% of the population gets its drinking water from groundwater resources. Groundwater also represents an important source of technical water in industry, as well as for energy production, food industry, agriculture, and tourism.

The geological structure of Slovenia is very diverse in terms of age (from Early Paleozoic to the present), lithology, and tectonics. Due to this geological diversity, the geological structure of the area is very complex [30], which makes the task of modeling groundwater properties even more challenging. The most abundant rock type is sedimentary, covering about 93% of the country, while igneous and metamorphic rocks cover much smaller areas, 3% and 4%, respectively [31]. Groundwater occurs in different geological structures. According to the type of porosity, aquifers can be divided into aquifers with intergranular, fissured, and karstic porosity (Figure 1). About 14% of the area is gravel/sandy deposits with intergranular porosity, 45% of the area is represented by karstic/fissured aquifers, and 35% of the area is represented by minor aquifers in porous or fissured formations with local and limited groundwater resources. Overlying strata or layers with essentially no significant groundwater resources comprise some 6% of the area [32].

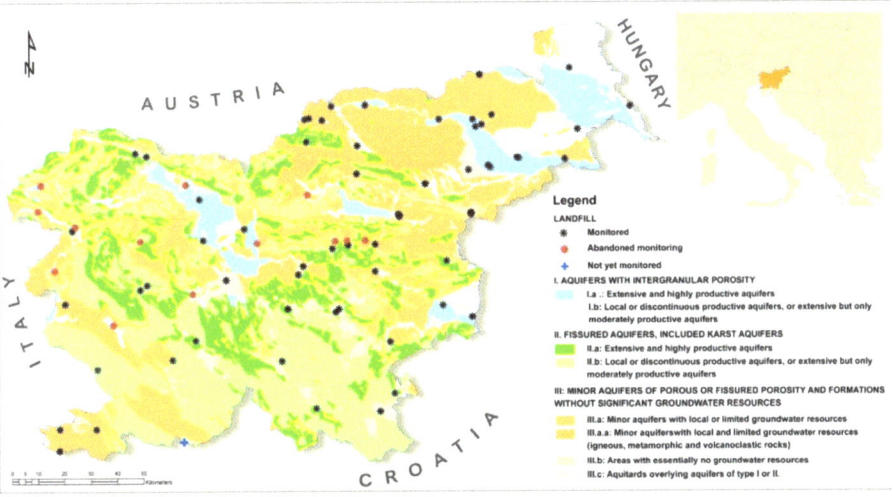

Figure 1. Study area and landfill locations.

2.2. Overview of Landfill Characteristics and Spatial Distribution

A total of 69 landfills were included in the analysis. Municipal solid waste type of landfill predominates (50 landfills or 72%) (Table 1), while seven (10%) landfills are inert waste type, 10 landfills are nonhazardous waste type (15%), and two landfills (3%) are hazardous waste type. In addition, 14 active landfills (10 municipal, one inert, one hazardous, and two nonhazardous landfills), 15 closure phase landfills (12 municipal and three inert landfills), and 40 closed landfills (28 municipal, three inert, one hazardous, and eight nonhazardous landfills) are monitored. Additional information is given in the Supplementary Material S1 (sheet "class"), i.e., the size and engineering characteristics of the landfill body, the main hydrogeological characteristics in the designated area of the landfill, possible receptors for each landfill, and the suitability of the monitoring system. The impact assessment of the landfill on groundwater quality and the evaluation of the reliability of the assessment are also presented.

Table 1. Overview of number of landfills by type, status, and location of groundwater protection zones and groundwater dependent-ecosystems according to status of the landfills.

Type of aquifer	n	Waste Type				Status			GWPZ			GDE		
		Municipal	Inert	Nonhazardous	Hazard	Active	In closing	Closed	Active	In closing	Closed	Active	In closing	Closed
Intergranular aquifers	23	18	3	2	0	3	4	16	1	1	4	1	-	1
Fissured/karstic aquifers	23	16	3	3	1	5	6	12	2	1	1	-	4	3
Minor aquifers *	23	16	1	5	1	6	5	12	2	-	2	1	-	1
	69	50	7	10	2	14	15	40	5	2	7	2	4	5

GWPZ—groundwater protection zone; GDE—groundwater-dependent ecosystem; * minor aquifers of porous or fissured porosity and formations without significant groundwater resources.

2.2.1. Landfill Characteristics—Engineering Disposition of the Landfill

An adequate engineering disposition (R) required by the regulation in Slovenia has 23 landfills. This means that the sealing of the ground with a clayey layer and PEHD foil is provided, and the collection of the leachate by drainage system to retention basins or to treatment plants is on place. A partially settled (P) liner and/or drainage system has 34 landfills. Most of these landfills consist of an old and a new part of the landfill. The old parts are usually not underlined by impervious foil but only the clayey layer lies in the base of the waste. The newer parts of landfills are underlined by PEHD foil, or the waste is located exclusively on natural, poorly permeable sediments (clay and marl). In these landfills, leachate drainage is partially adequate, due to either an ineffective wastewater treatment plant or the old part of the landfill not having operational drainage, while the new part of the landfill does. Engineering characteristics of 12 landfills are unregulated (U). The waste is deposited directly on permeable layers (gravel and carbonate rock), the drainage system is not adequate, and the leachate infiltrates into the subsoil.

2.2.2. Spatial Distribution—Natural Characteristics of the Site of the Landfill

The natural characteristics of the site and downgradient area represent the primary factor for the extent of the landfill's impact on groundwater quality (permanent or temporary). Therefore, the primary analysis was conducted with the aid of a hydrogeological map (Figure 1) representing basic aquifer types [32]. Table 1 shows that 23 of 69 landfills are located on intergranular aquifers (three active), 23 (five active) are located on karstic/fissured aquifers, and 23 (six active) are located on minor aquifers with local and limited groundwater resources.

Moreover, 17 landfills are located in the areas of the most important aquifers (type I.a and II.a), which represents about 25% of all landfills in Slovenia. Their medium to high productivity and regional extent make the potential of these aquifers very important for drinking water supply, presently and in the future.

There are 14 landfills (five active) situated in water protection zones (GWPZs) (Table 1). Only one landfill (closed) is located directly in the groundwater-dependent ecosystem area (GDE), while 10 landfills extend into the GDE area via their "hydrogeologic target zones".

2.3. Source of Data

The dataset was obtained from the Slovenian Environment Agency (ARSO) within the Ministry of the Environment and Spatial Planning, including 82 landfills from the whole territory of Slovenia. Annual monitoring, which represents the basis to determine and assess the landfill leaching effect on groundwater, is conducted at 68 landfills, while monitoring is not yet performed for one landfill. Monitoring was abandoned at 13 landfills where it was demonstrated that pollution is not spreading to groundwater (Figure 1). Thus, the developed integrated methodology was tested at 68 landfills.

Data on the main hydrogeological characteristics of the landfill, the engineering characteristics of the landfill, and the results of groundwater chemical monitoring were obtained. Representative groundwater quality data were collected for 65 landfills in 2017. At three landfills, no monitoring was conducted in this year. For these landfills, monitoring data from the most recent available year until 2017 were used.

2.4. Groundwater Monitoring at the Landfill Area
2.4.1. Groundwater Monitoring Network and Sampling

The monitoring network at the landfills consisted of a minimum of three monitoring points, of which at least two monitoring points were located downgradient and one monitoring point was located upgradient of the landfill (Figure 2). Groundwater chemical monitoring was conducted in the target hydrogeologic zone where contamination could be expected due to indirect or direct discharge of contaminants from a source of contamination to groundwater.

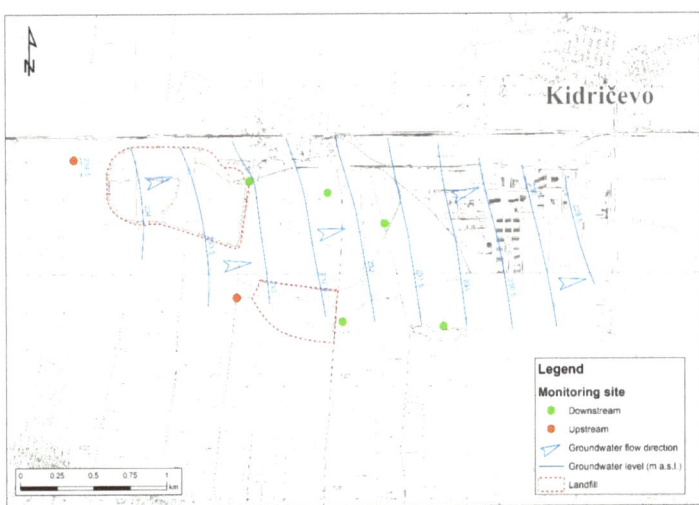

Figure 2. Example of the monitoring site positions at landfill.

The sampling set on karst aquifers consisted mostly of springs and to a lesser extent boreholes. The distances between monitoring points were in most cases greater than 10 km.

However, on intergranular aquifers and aquifers with local and limited groundwater resources the sampling set consisted of boreholes, private wells and pumping stations.

Annual monitoring of the landfill impact on groundwater quality was carried out by various accredited laboratories in Slovenia, such as the National Laboratory for Health, Environment and Food, Eurofins Erico Slovenija d.o.o., JP Vodovod Kanalizacija Snaga d.o.o., Talum Inštitut d.o.o., Tab-Ipm Logistika, Plastika In Storitve d.o.o., and Regionalni tehnološki center Zasavje d.o.o. Groundwater sampling in areas where karst porosity prevails was performed four times per year, one in each season. In the areas with fissured and intergranular porosity, groundwater sampling was performed twice per year with the sampling interval of at least 3 months and no longer than 6 months. In general, sampling campaigns were conducted under hydrogeological baseflow conditions (predominant discharge from aquifer storage).

Sampling procedures, transport, and storage of the groundwater samples were performed in accordance with ISO standards (SIST ISO 5677-11:1996; SIST ISO 5677-03:1996; SIST ISO 5677-6:1996).

2.4.2. Chemical Analysis

The groundwater samples were analyzed at the accredited laboratories in Slovenia mentioned in the previous section.

In our study, field parameters (temperature, pH, electrical conductivity, dissolved oxygen, and redox potential), inorganic parameters (Na^+, K^+, Ca^{2+}, Mg^{2+}, NH_4^+, HCO_3^-, SO_4^{2-}, Cl^-, NO_3^-, F^-, NO_2^-, and PO_4^{3-}), microelements (B, Al, As, Sb, Cu, Ba, Be, Zn, Cd, Co, Sn, Cr total, Cr^{6+}, Mo, Mn, Ni, Se, Ag, Pb, Tl, Ti, Te, V, Fe, and Hg), and organic compounds (total organic carbon (TOC), adsorbable organic halides (AOX), volatile organic compounds (VOCs), aromatic hydrocarbons (BTEX), triazine, and organochlorine pesticides) in groundwater were recorded. For microelements, dissolved species in groundwater were measured. The organic compounds included in the monitoring were selected on the basis of the type of waste and the results of the groundwater zero status. Table 2 shows the analytical procedure used for each chemical parameter and measurement uncertainties.

Table 2. Analytical procedure and measurement uncertainties for each chemical parameter in groundwater.

Parameter	Analytical Procedure	Measurement Uncertainty
Temperature	SIST DIN 38404-4:2000	0.3 °C
pH	ISO 10523: 2008	0.12
Electrical conductivity	EN 27888:1993	2.0%
Dissolved oxygen	ISO 17289.:2014	10%
Redox potential	DIN 38404-C6: 1984	10%
Na^+, K^+, Ca^{2+}, and Mg^{2+}	EN ISO 14911: 1999	10–19%
SO_4^{2-}, Cl^-, and NO_3^-	ISO 10304-1: 2007	11–15%
HCO_3^-	EN ISO 9963-1: 1995	10%
NH_4^+	ISO 11732: 2005	13%
NO_2^-	ISO 13395: 1996	13%
PO_4^{3-}	ISO 15681-2: 2018	10%
Microelements	ISO 17294-2: 2016	6–16%
AOX	ISO 9562: 2004	26%
TOC	ISO 8245: 1999	11%
VOCs and BTEX	EN ISO 15680: 2003	30%
Pesticides	EN ISO 11,369 modif.: 1997	4–17%
PAO	EN ISO 17,993 modif.: 2002	20–29%

2.4.3. Evaluation of Chemical Data

Chemical groundwater monitoring datasets were presented up to the concentration values of the lower limit of quantification (LOQ). LOQ values varied for individual parameters on a case-by-case basis, between individual measurement series, and by laboratory. The highest LOQ values of each parameter were used for analysis. All numerical results below the selected value of LOQ were considered results at LOQ. For data at LOQ, 50% of the LOQ values were used for further data processing in accordance with the recommendation in Annex 3 of the National Regulation on groundwater status.

The impact of landfills on groundwater was firstly assessed by comparing the annual mean values of each parameter in the groundwater at monitoring points upstream and downstream of the landfill, according to the following equation:

$$X_{mean,DMP} \geq X_{mean,UMP} \qquad (1)$$

where DMP is the annual mean value of each parameter at the downgradient monitoring point, and UMP is the annual mean value of each parameter at the upgradient monitoring point. If the DMP value was greater than or equal to the UMP value, a value of "1" was assigned, indicating that the landfill had a potential impact on groundwater quality. Otherwise, the value "0" was used.

For the parameters with the value of "1", the exceedance of the values for each parameter was further evaluated according to the following equation:

$$\left(\left(X_{mean,DMP} - \frac{X_{mean,DMP} \cdot MU}{100}\right) - \left(X_{mean,UMP} - \frac{X_{mean,UMP} \cdot MU}{100}\right)\right) \cdot \frac{100}{X_{mean,UMP}} \qquad (2)$$

where MU is the measurement uncertainty for each parameter.

According to the calculations, only the chemical parameters that determined a significant impact on the groundwater quality of the landfill were considered. Calculated values $\geq 100\%$ indicated that the measured values of each parameter at the downstream monitoring point were significantly higher than the measured values of the parameter at the upstream monitoring point. The parameter indicated the influence of the landfill on the status of the groundwater and was further considered as an indicative parameter. Calculated values were between $\geq 0\%$ and 100% indicated that the measured content of the respective parameter at the downstream monitoring point was not significantly higher than the measured parameter values at the upstream monitoring point. The assessment of the impact of the landfill on the groundwater status for the selected parameter was not reliable and was not used to assess the impact of the landfill. Calculated values 0% indicated that the measured values of each parameter at the downstream monitoring point were less than the values at the upstream monitoring point. In this case, the groundwater at the upstream monitoring point was already polluted compared to the situation at the downstream monitoring point. Thus, assessment of the landfill's influence on the groundwater status was not possible.

2.5. Assessment of Landfill Impacts on Groundwater

2.5.1. Conceptual Model

The assessment of landfill impacts on groundwater quality status proposed herein is based on a conceptual model of the landfill, which is intended to clearly define the hydrogeologic conditions in the landfill area (Figure 3). The conceptual model summarizes and evaluates all available data related to the landfill, as well as those obtained during the monitoring period. At the same time, it also identifies the weaknesses and uncertainties related to the available data. In order to carry out an impact assessment, as well as optimize decisions for future pollution management (e.g., monitoring and preventive measures), a conceptual model must answer three specific questions: (1) Do the natural conditions and engineering characteristics of the landfill allow the leaching of pollutants into groundwater; if yes, to what extent, and is the pollutant leaching direct or indirect? (2) Does existing monitoring allow the control of potential pollutant leaching from the landfill area? Is the monitoring network adequate or does it need to be optimized? (3) Is it possible to confirm the environmental impact and estimate its extent?

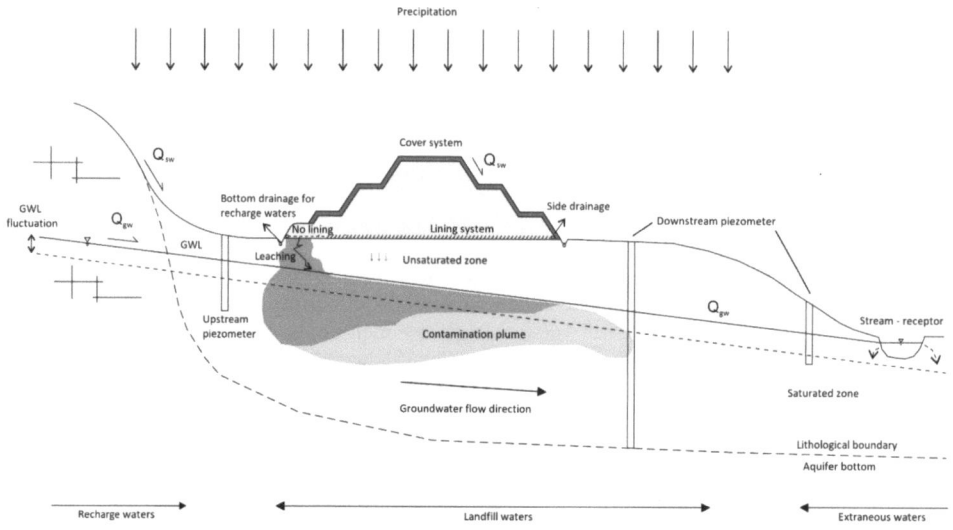

Figure 3. Conceptual model of the landfill.

These questions can be answered if the appropriate information is available, which considers (1) the engineering characteristics of the landfill, including the nature of the pollutants, (2) the natural (e.g., hydrogeological) conditions of the landfill, on the basis of which the risk potential is determined, together with the quantities of leachate and the concentration of the pollutants, (3) possible contamination transport pathways and the presence of indicative contaminants in the groundwater, and (4) potentially affected receptors.

2.5.2. Engineering Characteristics of the Landfill

An important factor for potential groundwater contamination is the engineering characteristics of the landfill, including the landfill liner (lining system), drainage performance and position of drainages, leachate collection, and slope stability. All these conditions are essential to understand potential emissions of pollutant into the environment. For example, if the sealing layer is damaged, direct leaching to groundwater may occur. The quality of the sealing layers and the cover is important to ensure effective cover and to ensure the least possible emissions into surface and groundwater. At least around the landfill body, the natural base of the landfill must be geologically and hydrogeologically uniform and should have proper geological characteristics to ensure protection against contamination of the soil, surface water, and groundwater. The average water permeability of the landfill cover layers must be less than 1×10^{-9} m/s (Decree on the landfill of waste (Official Gazette of RS, no. 10/14, 54/15, 36/16, 37/18 and 13/21)). If the permeability of the soil is higher, artificial sealing layers (e.g., HDPE foil) are also installed to ensure protection against contamination. The landfill body and its subsoil must be stable in the long term so that possible deformations will not affect the landfill sealing, the leachate and rainwater drainage system, or landfill degassing system. Only constant, unchanged conditions enable controlling the water cycle and the pollutant mass balance, which is the basis for a successful impact and risk assessment.

2.5.3. Hydrogeological Conditions

Karstic/fissured aquifers (Figure 1) are heterogeneous formations; hence, it is very difficult to predict groundwater hydraulics and mass transport [33,34]. Another problem is a significant variability in the discharge behavior of karst water under different hydrologic conditions. In some karst fissures and channels, water can drain very rapidly, while, in others, it can be retained much longer [35–37]. The difference in groundwater levels has a significant impact on the direction and the travel time of groundwater, as well as on the possibility of dilution and storage of contaminants in the subsurface. From this perspective, landfill sites on karst aquifers should be treated differently than those on intergranular aquifers. Aquifers with karstic/fissured porosity are very vulnerable. Therefore, new landfills in karstic areas are prohibited in Slovenia. The risks posed by landfills in karstic areas are associated with the heterogeneity of the hydrogeological characteristics, such as highly variable spatial distribution of (1) fractures in the bedrock (2), permeability of aquifer, and (3) groundwater flow velocity and direction. In the region of highly fractured or karstified bedrock and variable (spatially and temporarily) groundwater flow conditions, there is a high risk that pollution from the landfill will spread to a wider area.

The recognition and the evaluation of groundwater flow hydraulic properties in intergranular aquifers (Figure 1) are relatively straightforward. In such areas, with an appropriate monitoring network and under well-known boundary conditions, the contamination plume downgradient from the landfill and outside of the monitoring network can be followed using analytical methods and simulations. The greatest risk for contamination spreading is posed by surface water, which can drain relatively large quantities (volumes) of groundwater. In porous aquifers, interactions between groundwater and surface water often occur, resulting in the direct impact of contamination on local ecosystems or even on groundwater sources used for the drinking water supply.

The movement of contaminants in groundwater depends on the natural attenuation of the soil and natural degradation processes in the aquifer. Natural attenuation is the result of various physical, chemical, and biological processes and can reduce the mass, toxicity, mobility, volume, or concentrations of soil or groundwater contaminants without human intervention. These processes include biodegradation, dilution, sorption, evaporation, and radioactive decay. Natural attenuation occurs in all contaminated areas, but the rate of the degradation process depends on certain conditions in the soil. If such conditions are not significantly present, contaminant degradation is not effective nor comprehensive [38,39].

Determination of hydrogeological parameters (hydraulic conductivity, porosity, etc.) of the aquifer is important for the proper control and prediction of contaminant transport including plume spreading. Here, the nature of contaminant emissions to groundwater is also an important factor. Several conditions can be interpreting regarding the contamination source (landfill body) position and groundwater level. The direct source of contamination is when the landfill body (partially or wholly) is permanently or occasionally in a saturated zone. When the landfill body is permanently in unsaturated zone, this can be interpreted as an indirect contamination source [40].

2.5.4. Potential Pathway of the Pollution

When considering the contamination pathways in groundwater, it is important to consider the horizontal pathway along the main groundwater flow direction. The groundwater flow regime is a key parameter that allows indicating the relative length of the path of the contaminant, the travel time, and the possible interaction with surface water.

Quality of the monitoring network is a crucial factor to determine the extent of the contamination plume estimation and, consequently, of the risk assessment. A proper monitoring network enables an early warning system and long-term forecast. It includes several observation points along the groundwater flow path. A monitoring site upgradient from the landfill can be used to determine potential changes in natural background levels and to define the extent of the impact. Analysis of leachate represents the range and mass of pollutants at the emission point (point of compliance POC 0). The monitoring site below the landfill (POC 1) in the contaminant plume at the entrance to the groundwater is the key monitoring point, revealing the pressure of the landfill to groundwater. POC 2 is the early warning monitoring site downstream between the landfill and the monitoring site POC 3 at the receptor (spring, well, etc.) [41].

2.5.5. Potential Receptors Affected by the Pollution

Evaluation of the actual landfill impact is based on the determination of potential receptors and their position. In this study, the receptors are presented as sensitive areas on which the potential contamination spreading from landfill may have a strong impact. Such areas are the water protection zones of drinking water and areas with the presence of the groundwater-dependent ecosystems.

Groundwater protection zones are delimited according to the methodology provided in national regulations (Rules on criteria for the designation of a water protection zone (Official Gazette RS, no. 64/04, 5/06, 58/11 and 15/16)). The purpose of groundwater protection zones is to prevent and restrict points and diffuse sources of pollution that may affect drinking water quality [42]. Protective measures, prohibitions, and restrictions apply to construction and other land uses. In Slovenia, approximately 17% of the territory is under a water protection zone regime [43].

As an integral part of the groundwater body status assessment in Slovenia, areas with groundwater-dependent ecosystems (GDEs) have been identified on the basis of hydrological and hydrogeological conditions within the territory of Slovenia. GDEs include ecosystems in groundwater, terrestrial ecosystems dependent on groundwater, and aquatic ecosystems in surface waters dependent on groundwater [44,45]. They require constant or occasional contact with groundwater to sustain communities of diverse animal and plant species, ecological processes, and ecosystem services [45]. Consequently, they are a good indicator of the status of groundwater bodies and play an important role in their assessment [46,47]. In total, GDEs cover about 28% of the Slovenian territory.

2.6. Synoptic Risk Assessment

The methodology for classifying landfills into priority classes was developed according to the criteria of individual components of the conceptual model and the chemical characteristics of groundwater (Table 3). According to the presented criteria, we made a decision tree for each type of aquifer (intergranular aquifers, fissured/karstic aquifers, and minor aquifers of porous or fissured porosity and formations without significant groundwater resources), on the basis of which we classified the landfill into an individual priority class, as shown in the Supplementary Material S2.

Table 3. Criteria of hydrogeological and engineering characteristics for classifying landfills into three priority classes.

	YES	NO	PARTLY
Engineering Characteristics	Regulated (R)—see Section 2.2.1.	Unregulated (U)—see Section 2.2.1.	Partly regulated (P)—see Section 2.2.1.
Input of Contamination	Direct input of contamination—see Section 2.5.3.	Indirect input of contamination—see Section 2.5.3.	-
GDE, GWPZ	Present	Not present	Present on target hydrogeological zone
Chemical Analysis	Calculated values ≥100%—see Section 2.4.3.	Calculated values 0%—see Section 2.4.3.	Calculated values between ≥0% and 100%—see Section 2.4.3.
Net of Sampling Points	Appropriate	Inappropriate	-

Classification of landfills was based on the assumption of setting priorities of protection against impacts on groundwater quality. In terms of the groundwater risk pollution, landfills were classified into three priority classes: (1) landfills with a significant impact on local groundwater quality and a potential impact on quality of the groundwater body. These are landfills where contamination may spread with groundwater to the wider area, posing a risk to potential drinking water supplies and/or groundwater-dependent ecosystems downgradient from the landfill site; (2) landfills with less significant impacts on groundwater, either due to favorable natural attenuation processes or due to adequate engineering performance of the landfill itself. These are landfills where the natural attenuation of the subsurface is assumed to be sufficiently efficient to prevent the spread of contaminants with groundwater flow. Impact on groundwater quality is limited to the local area of the landfill; (3) landfills with insignificant impact or no impact on groundwater quality. These landfills and their impact areas are located in the areas of limited groundwater resources or even in the areas without groundwater. Pollutants from the landfill cannot be spread by groundwater flow. There is no risk to groundwater resources or to groundwater-dependent ecosystems.

3. Results and Discussion

3.1. Indicative Parameters of Impacts on Groundwater

Indicative parameters for 68 landfills were determined through the data processing presented in Section 2.4.3. According to this analysis, the indicative parameters indicating the impact of the landfill on groundwater quality status (calculated values $\geq 100\%$) were determined for each landfill, (Figure 1).

Those parameters that occur in at least 10% ($\pm 1\%$) of all landfills were included in further analysis. We identified 43 indicative parameters, which were classified into three groups, namely, major parameters, microelements, and organic compounds. We assigned the number of landfills according to type of landfill to each identified indicative parameter. The classification is shown in Figure 4 and Table 3.

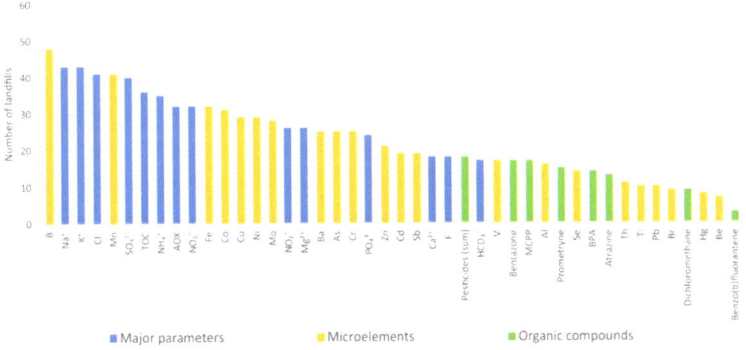

Figure 4. Frequencies of determined parameters.

Figure 4 shows the frequencies of the indicative parameters indicating the impact of the landfill on groundwater quality. The highest frequency was obtained for boron (47) under the group of microelements, followed by major parameters K^+, Na^+, Cl^-, SO_4^{2-}, NH_4^+, and NO_3^- (46–31), and some organic compounds TOC (36) and AOX (32). Most of the microelements were determined as indicative parameters from 19 to 31 times. Among them, Mn corresponded with 41 determinations and Fe corresponded with 32 determinations. However, some of them were also detected in smaller frequencies (7–17 times). Organic compounds were represented by seven parameters in frequencies ranging from 13 to 18 instances. In this group, we most frequently recognized total pesticides.

We performed a detailed analysis by individual group of parameters and type of landfill. Table 4 shows the frequencies of determination of indicative parameters by group of parameters and type of landfills.

Table 4. Identification of indicative parameters by group of parameters and type of landfill.

		Municipal	Inert	Non-Hazardous	Hazard	Total (n)	Total (%)
	No. of Landfills	49	7	10	2	68	
Major parameters	K^+	32	3	6	2	43	63
	Na^+	30	4	7	2	43	63
	Cl^-	31	2	6	2	41	60
	SO_4^{2-}	28	4	6	1	39	57
	NH_4^+	28	0	6	1	35	51
	NO_3^-	23	3	5	0	31	46
	NO_2^-	23	0	3	0	26	38
	Mg^{2+}	17	2	5	1	25	37
	PO_4^{3-}	19	1	4	0	24	35
	Ca^{2+}	13	3	2	0	18	26
	F^-	10	3	5	0	18	26
	HCO_3^-	14	2	1	0	17	25
Microelements	B	34	4	7	2	47	69
	Mn	31	3	6	1	41	60
	Fe	27	2	2	1	32	47
	Co	26	2	3	0	31	46
	Mo	22	5	4	0	31	46
	Cu	24	2	3	0	29	43
	Ni	25	2	2	0	29	43
	As	15	2	8	0	25	37
	Ba	17	2	5	1	25	37
	Cr	16	3	3	1	23	34
	Zn	17	0	4	0	21	31
	Sb	13	2	4	1	20	29
	Cd	13	2	3	1	19	28
	V	13	2	2	0	17	25
	Al	13	0	2	1	16	24
	Se	8	1	4	1	14	21
	Tl	10	0	1	0	11	16
	Pb	9	0	1	0	10	15
	Br	6	0	2	1	9	13
	Hg	5	1	2	0	8	12
	Be	5	0	1	1	7	10
	Ti	4	2	0	1	7	10
Organic compounds	TOC	31	2	2	1	36	53
	AOX	25	3	3	1	32	47
	Pesticides (total)	16	0	1	1	18	26
	Bentazone	16	0	0	1	17	25
	MCPP	16	0	0	1	17	25
	Prometryne	17	0	0	0	17	25
	DEET	16	0	0	0	16	24
	Atrazine	11	1	1	0	13	19
	BPA	11	0	2	0	13	19

K^+, Cl^-, TOC, and Na^+ were recognized as the most indicative parameters under the group major parameters and organic compounds in the municipal landfills group, as they, among other major ions, control the mineralization processes in the landfill body and can occur in different compound combinations as of geogenic (natural) origin or as a result of leaching from landfilled waste materials (of anthropogenic origin). Among them, NO_3^-, NO_2^-, and NH_4^+ were also frequently recognized as indicative parameters in the literature [8].

In the group of microelements, Fe and Mn were typically determined as indicative parameters as expected [8,24], since both occur in areas where the reduction conditions downstream of the landfill are significant [3]. The metals B, Co, and Ni were also detected at more than 50% of the landfills, as expected [48]. Organic compounds were detected as typical contaminants in municipal landfills. Prometryne was most frequently detected as an indicative parameter, followed by pesticides (total) MCPP, bentazone, and DEET. Atrazine and BPA were less frequently determined. The presence of pesticides in groundwater at landfill sites is a complex issue owing to the possibility that these substances may be present simultaneously and may originate from different sources. Pesticides were already present in groundwater at the upstream sampling point as the result of their use on agricultural land. However, the groundwater in the landfill area was additionally loaded with pesticides due to the deposited landfill waste, as evidenced by the higher concentrations of pesticides in groundwater at downstream sampling points (Supplementary Material S1 (sheet "Main parameters at each landfill")).

Na^+, K^+, and SO_4^{2-} were identified as indicative parameters from the group major ions in the groups inert and nonhazardous landfills with fewer instances. Among microelements, Mn and B were identified as indicative parameters in more than 50% of the landfills in both groups. Inert landfills were also characterized by Mo (most frequent, 5/7) and Cr (3/7), while nonhazardous landfills were characterized by As (most frequent, 8/10) and Ba, as well as Cl^- and NH_4^+. Inert landfills mostly constituted fly ash, slag, and construction waste from which nonhazardous inorganic compounds (sulfate and sodium) and metals were leached. Organic compounds were detected individually, and usually in a single instance.

Na^+, K^+, Cl^-, and B were identified as indicative parameters at two hazardous landfills.

3.2. Synoptic Risk Assessment

Using the methodology presented in Section 2.5, the classification of landfills was carried out (Table 5, Figure 5 and Supplementary Material S1 (sheet "class")).

Table 5. Classification of landfills based on the conceptual model method.

	Aquifer Type	Class 1	Class 2	Class 3
	Significant impact	+	o	−
	GWPZ in/or GDE presence	+	+	−
	Technical suitability	−	−	o
I.a; I.b	Aquifers with intergranular porosity	8	11	4
II.a; II.b	Fissured aquifers, including karst aquifers	8	8	7
III.a; III.c	Minor aquifers of porous or fissured porosity and formations without significant groundwater resources	8	12	3
		24	31	14

+—yes; o—partially; −—no; GWPZ—groundwater protection zone; GDE—groundwater-dependent ecosystem.

A total of 24 landfills that had a significantly pronounced impact on groundwater and could also be reflected in the chemical status of the groundwater body were classified as Priority Class 1. Among those, eight landfills were located in the area of aquifers where intergranular porosity (type I.a and I.b) was predominant, eight landfills were located in aquifers with predominantly karstic/fissured porosity (type II.a and II.b), and eight landfills were located in the area of minor aquifers with local and limited groundwater sources (III.a). All landfills located directly in a groundwater protection zone, where direct inflow and rapid transport of pollutants were identified, and which were also engineering unregulated, were classified as Class 1. For most of these landfills, the impact on groundwater was clear. In addition, landfills were included in Priority Class 1, where the waste is deposited directly in an abandoned gravel pit without sealing and under unfavorable hydrogeological conditions (highly permeable aquifer and high groundwater flow velocity).

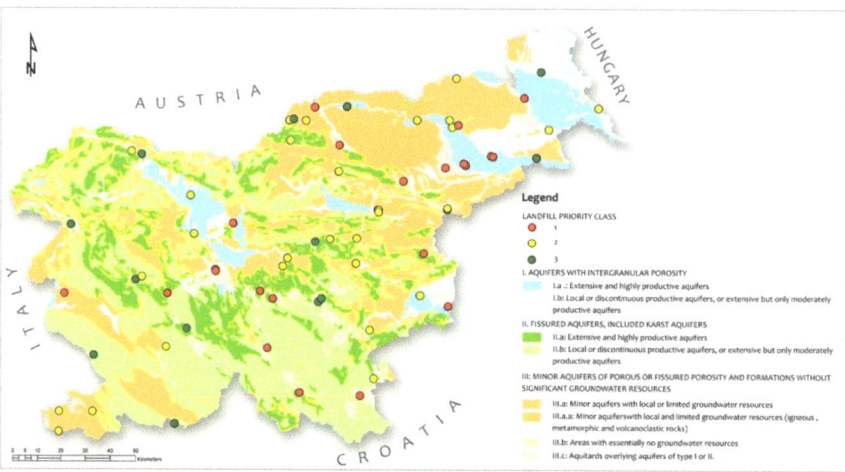

Figure 5. Distribution of priority classes for each landfill.

A total of 31 landfills with less significant impacts were classified with Priority Class 2 due to favorable natural attenuation below the landfill or landfills with suitable engineering characteristics. Of these, 11 landfills were located in the vicinity of aquifers dominated by intergranular porosity (types I.a and I.b), eight landfills were located in the area of aquifers dominated by karstic/fissured porosity (type II.a and II.b), and 10 landfills were located in the area of minor aquifers with local and limited groundwater resources (III.a). Two landfills were located in the area of poorly permeable roof layers (type III.c).

Class 2 also included landfills where the impact of the landfill was not as significant due to groundwater pressures in the affected aquifer, but which were located in groundwater protection zones. Class 2 also included landfills that, owing to their engineering characteristics or the prevailing natural conditions, allowed significant inputs of pollutants into the groundwater. The contamination plume was also not expected to affect the wider environment.

Landfills located in areas of karstic/fissured porosity were considered separately due to uncertainty in determining the groundwater flow direction and its velocity. An additional issue was the distance between monitoring points, which, in most cases, was greater than 10 km. At these distances, there were known and unknown surface sources that may have a greater impact on groundwater status than the landfill; therefore, it was not possible to reliably assess the potential impact of the landfill on the status of the groundwater body.

A total of 14 landfills with little or no impact on groundwater quality were included in Priority Class 3. Most of the landfills (seven) were on aquifers dominated by karstic/fissured porosity, four landfills were on aquifers dominated by intergranular porosity, and three were located on minor aquifers with local and limited groundwater resources. The main criteria for the classification of these landfills were as follows: the landfill was small to medium-sized, where large amounts of leachate were not expected, and the engineering characteristics of the landfill were at least partially regulated. This class also included landfills where expected groundwater pressures were low or absent, and the natural conditions were sufficiently good, as well as if the contamination plume could be followed through monitoring to predict the spatial and temporal distribution of contaminants.

4. Conclusions

This study presented a developed method of synoptic risk assessment based on a conceptual model of the landfill and the results of the groundwater chemical monitoring, which was tested on 69 landfills in Slovenia.

The study confirmed that majority of landfills in Slovenia can affect groundwater quality. On average, these impacts are small to negligible; however, in some cases, they can be significant and can also affect the quality of the groundwater body.

Analysis of chemical groundwater monitoring showed that B is the most often determined indicative parameter in groundwater, followed by Mn-total, Fe-total, K^+, Na^+, Cl^-, and SO_4^{2-}. The least frequently detected indicative parameters appeared in the group organic compounds, with the most frequently detected TOC and AOX.

Landfills differed significantly depending on the wastes they receive. For inert landfills, Mo occurred as an indicative parameter in more than 70% of cases, while As was most frequently detected in non-hazardous landfills. F^-, Mg^{2+}, and NO^{3-} were only determined at a frequency rate of more than 50% in non-hazardous landfills, while AOX, TOC, Co, Fe-total, and Ni were only present at more than 50% in the municipal landfills. As expected, organic compounds were identified as typical contaminants mainly in municipal landfills.

The landfill classification used showed that a total of 24 landfills had a significantly pronounced impact on groundwater and may also be reflected in the chemical status of the groundwater body; accordingly, they were classified in Priority Class 1. These landfills should be given the greatest attention with an emphasis on protection measures. For these landfills, additional engineering measures are required to reduce the input of contaminants more effectively from the landfill into the groundwater.

A total of 31 landfills had less significant impacts due to favorable natural attenuation below the landfill or landfills with suitable engineering characteristics; accordingly, they belonged to Priority Class 2. For these landfills, appropriate measures need to be prepared. The main measure is to ensure reliable monitoring, while no other technical measures are expected except those that are mandatory.

A total of 14 landfills had little or no impact on groundwater quality and, accordingly, were included in Priority Class 3. These landfills are not expected to require additional measures beyond the basic landfill measures in place. There is no risk to groundwater sources or groundwater-dependent ecosystems.

The classification of landfill impacts based on the synoptic risk assessment presented in our study was found to be effective in assessing the environmental impact of landfills, and it can also be used as a guideline in other parts of the world. Annual monitoring enables landfill managers to control the efficiency of the protection measures and improve the program of measures if needed. The survey of monitoring results of all landfills enables permitting and managing authority to improve operational programs and action plans. This relies on a synoptic risk assessment which enables specifying the priorities and milestones of needed actions. As can be seen from the Water Framework Directive, preparation of the program will consider the principles aimed at preventing and limiting the discharge of substances.

Further work will focus on a statistical evaluation of the indicative parameters, as it is necessary to identify concentration trends and work on the possible concentration range of each indicative parameter. This information will be very important to determine whether the situation at individual landfills is deteriorating, improving, or even stagnating. This is also the basis for planning additional protection measures in the future.

Supplementary Materials: The following supporting information can be downloaded at: https://www.mdpi.com/article/10.3390/en15145150/s1. S1: Excel file with 2 sheets named "class" and "Main parameters at each landfil". S2: Word file named decission tree.

Author Contributions: Conceptualization, J.P.; Formal analysis, K.K.; Investigation, S.C.; Methodology, L.S.; Writing—original draft, S.C., L.S. and N.M.; writing—review and editing, S.C., L.S. and K.K.; visualization, K.K. and L.S. All authors have read and agreed to the published version of the manuscript.

Funding: This research was funded by the Slovenian Research Agency (ARRS), Research Program Groundwater and Geochemistry No. P1-0020.

Institutional Review Board Statement: Not applicable.

Informed Consent Statement: Not applicable.

Data Availability Statement: The data presented in this study are available on request from the corresponding author or from the Slovenian Environment Agency (Ministry of Environment and Spatial Planning).

Acknowledgments: The authors would like to thank the Slovenian Environment Agency (Ministry of Environment and Spatial Planning) for data sharing.

Conflicts of Interest: The authors declare no conflict of interest.

References

1. Bhalla, B.; Saini, M.S.; Jha, M.K. Effect of age and seasonal variations on leachate characteristics of municipal solid waste landfill. *Int. J. Res. Eng. Technol.* **2013**, *2*, 223–232. Available online: http://www.ijret.org (accessed on 18 May 2022).
2. Abiriga, D.; Vestgarden, L.S.; Klempe, H. Groundwater contamination from a municipal landfill: Effect of age, landfill closure, and season on groundwater chemistry. *Sci. Total Environ.* **2020**, *737*, 140307. [CrossRef]
3. Abiriga, D.; Vestgarden, L.S.; Klempe, H. Long-term redox conditions in a landfill-leachate-contaminated groundwater. *Sci. Total Environ.* **2021**, *755*, 143725. [CrossRef] [PubMed]
4. Butt, T.E.; Lockley, E.; Oduyemi, K.O. Risk assessment of landfill disposal sites—State of the art. *Waste Manag.* **2008**, *28*, 952–964. [CrossRef]
5. Castañeda, S.S.; Sucgang, R.J.; Almoneda, R.V.; Mendoza, N.D.S.; David, C.P.C. Environmental isotopes and major ions for tracing leachate contamination from a municipal landfill in Metro Manila, Philippines. *J. Environ. Radioact.* **2012**, *110*, 30–37. [CrossRef] [PubMed]
6. Christensen, T.H.; Kjeldsen, P.; Bjerg, P.L.; Jensen, D.L.; Christensen, J.B.; Baun, A.; Albrechtsen, H.; Heron, G. Biogeochemistry of landfill leachate plumes. *Appl. Geochem.* **2001**, *16*, 659–718. [CrossRef]
7. Samadder, S.R.; Prabhakar, R.; Khan, D.; Kishan, D.; Chauhan, M.S. Analysis of the contaminants released from municipal solid waste landfill site: A case study. *Sci. Total Environ.* **2017**, *580*, 593–601. [CrossRef]
8. Preziosi, E.; Frollini, E.; Zoppini, A.; Ghergo, S.; Melita, M.; Parrone, D.; Rossi, D.; Amalfitano, S. Disentangling natural and anthropogenic impacts on groundwater by hydrogeochemical, isotopic and microbiological data: Hints from a municipal solid waste landfill. *Waste Manag.* **2019**, *84*, 245–255. [CrossRef]
9. Kjeldsen, P.; Barlaz, M.A.; Rooker, A.P.; Baun, A.; Ledin, A.; Christensen, T.H. Present and long-term composition of MSW landfill leachate: A review. *Crit. Rev. Environ. Sci. Technol.* **2002**, *32*, 297–336. [CrossRef]
10. Bjerg, P.L.; Ruegge, K.; Pedersen, J.K.; Christensen, T.H. Distribution of redox-sensitive groundwater quality parameters downgradient of a landfill (Grindsted, Denmark). *Environ. Sci. Technol.* **1995**, *29*, 1387–1394. [CrossRef]
11. Hussein, M.; Yoneda, K.; Zaki, Z.M.; Amir, A. Leachate characterizations and pollution indices of active and closed unlined landfills in Malaysia. *Environ. Nanotechnol. Monit. Manag.* **2019**, *12*, 100232. [CrossRef]
12. Ančić, M.; Huđek, A.; Rihtarić, I.; Cazar, M.; Bačun-Družina, V.; Kopjar, N.; Durgo, K. PHYSICO chemical properties and toxicological effect of landfill groundwaters and leachates. *Chemosphere* **2020**, *238*, 124574. [CrossRef] [PubMed]
13. Baettker, E.C.; Kozak, C.; Knapik, H.G.; Aisse, M.M. Applicability of conventional and non-conventional parameters for municipal landfill leachate characterization. *Chemosphere* **2020**, *251*, 126414. [CrossRef] [PubMed]
14. Mishra, S.; Tiwary, D.; Ohri, A.; Agnihotri, A.K. Impact of Municipal Solid Waste Landfill leachate on groundwater quality in Varanasi, India. *Groundw. Sustain. Dev.* **2019**, *9*, 100230. [CrossRef]
15. Kapelewska, J.; Kotowska, U.; Karpińska, J.; Astel, A.; Zieliński, P.; Suchta, J.; Algrzym, K. Water pollution indicators and chemometric expertise for the assessment of the impact of municipal solid waste landfills on groundwater located in their area. *Chem. Eng. J.* **2019**, *359*, 790–800. [CrossRef]
16. Chidichimo, F.; De Biase, M.; Straface, S. Groundwater pollution assessment in landfill areas: Is it only about the leachate? *Waste Manag.* **2020**, *102*, 655–666. [CrossRef]
17. Liu, H.; Liang, Y.; Zhang, D.; Wang, C.; Liang, H.; Cai, H. Impact of MSW landfill on the environmental contamination of phthalate esters. *Waste Manag.* **2010**, *30*, 1569–1576. [CrossRef]
18. Peng, X.; Ou, W.; Wang, C.; Wang, Z.; Huang, Q.; Jin, J.; Tan, J. Occurrence and ecological potential of pharmaceuticals and personal care products in groundwater and reservoirs in the vicinity of municipal landfills in China. *Sci. Total Environ.* **2014**, *490*, 889–898. [CrossRef]
19. Kumar, D.; Alappat, B.J. Analysis of leachate pollution index and formulation of sub-leachate pollution indices. *Waste Manag. Res.* **2005**, *23*, 230–239. [CrossRef]
20. Chakraborty, S.; Kumar, R.N. Assessment of groundwater quality at a MSW landfill site using standard and AHP based water quality index: A case study from Ranchi, Jharkhand, India. *Environ. Monit. Assess.* **2016**, *188*, 335. [CrossRef]
21. Talalaj, I.A. Adaptation of water quality index (WQI) for groundwater quality assessment near the landfill site. *J. Water Chem. Technol.* **2014**, *36*, 144–151. [CrossRef]
22. Talalaj, I.A.; Biedka, P. Use of the landfill water pollution index (LWPI) for groundwater quality assessment near the landfill sites. *Environ. Sci. Pollut. Res.* **2016**, *23*, 24601–24613. [CrossRef] [PubMed]

23. Zhang, Y.; Hou, K.; Qian, H. Water quality assessment using comprehensive water quality index and modified Nemerow index method: A case study of Jinghui Canal, North China. In *IOP Conference Series: Earth and Environmental Science*; IOP Publishing: Bristol, UK, 2020; Volume 467, p. 012125.
24. Stefania, G.A.; Zanotti, C.; Bonomi, T.; Fumagalli, L.; Rotiroti, M. Determination of trigger levels for groundwater quality in landfills located in historically human-impacted areas. *Waste Manag.* **2018**, *75*, 400–406. [CrossRef]
25. Preziosi, E.; Giuliano, G.; Vivona, R. Natural background levels and threshold values derivation for naturally As, V and F rich groundwater bodies: A methodological case study in Central Italy. *Environ. Earth Sci.* **2010**, *61*, 885–897. [CrossRef]
26. Ducci, D.; de Melo, M.T.C.; Preziosi, E.; Sellerino, M.; Parrone, D.; Ribeiro, L. Combining natural background levels (NBLs) assessment with indicator kriging analysis to improve groundwater quality data interpretation and management. *Sci. Total Environ.* **2016**, *569*, 569–584. [CrossRef]
27. Serianz, L.; Cerar, S.; Šraj, M. Hydrogeochemical characterization and determination of natural background levels (NBL) in groundwater within the main lithological units in Slovenia. *Environ. Earth Sci.* **2020**, *79*, 1–17. [CrossRef]
28. Cerar, S.; Serianz, L.; Udovč, J.; Prestor, J.; Koren, K.; Lapanje, S. Analysis of groundwater monitoring programs and reports in the area of landfills for 2017: A synthetic risk assessment for the spread of groundwater pollution from landfills. *Rep. Geol. Surv. Slov.* **2018**.
29. SURS. Latest Data on Quantity of Waste in Slovenia. 2020. Available online: https://www.stat.si (accessed on 9 February 2021).
30. Pleničar, M.; Ogorelec, B.; Novak, M. *The Geology of Slovenia*; Geological Survey of Slovenia: Ljubljana, Slovenia, 2009.
31. Komac, M. Statistics of the Geological Map of Slovenia at scale 1: 250.000. *Geologija* **2005**, *48*, 117–126. [CrossRef]
32. Prestor, J.; Meglič, P.; Janža, M.; Bavec, M.; Komac, M. *Hydrogeological Map of Slovenia 1:250:000*; Geological Survey of Slovenia: Ljubljana, Slovenia, 2008.
33. Gabrovšek, F.; Knez, M.; Kogovšek, J.; Mihevc, A.; Mulec, J.; Perne, M.; Petrič, M.; Pipan, T.; Prelovšek, M.; Slabe, T.; et al. Development challenges in karst regions: Sustainable land use planning in the karst of Slovenia. *Carbonates Evaporites* **2011**, *26*, 365–380. [CrossRef]
34. Kresic, N. *Water in Karst*; McGraw Hill: New York, NY, USA, 2013.
35. Bonacci, O. *Karst Hydrology*; Springer: Berlin/Heidelberg, Germany, 1987; p. 184. [CrossRef]
36. Bonacci, O. Karst springs hydrographs as indicators of karst aquifers. *Hydrol. Sci. J./J. Des. Sci. Hydrol.* **1993**, *38*, 51–62. [CrossRef]
37. Petrič, M.; Kogovšek, J.; Ravbar, N. Effects of the Vadose Zone on Groundwater Flow and Solute Transport Characteristics in Mountainous Karst Aquifers—The Case of the Javorniki–Snežnik Massif (SW Slovenia). 2018, p. 47. Available online: https://ojs.zrc-sazu.si/carsologica/article/view/5144 (accessed on 8 July 2022).
38. Bedient, P.B.; Rifai, H.S.; Newell, C.J. *Ground Water Contamination—Transport and Remediation*; PTR Prentice-Hall, Inc.: Hoboken, NJ, USA, 1999.
39. Domenico, P.A.; Schwartz, F.W. *Physical and Chemical Hydrogeology*, 2nd ed.; John Wiley & Sons: Hoboken, NJ, USA, 1998; p. 528.
40. European Communities. Common Implementation Strategy for the Water Framework Directive (2000/60/EC): Guidance on Preventing or Limiting Direct and Indirect Inputs in the Context of the Groundwater Directive 2006/118/EC. Guidance Document No. 17. 2007. Available online: https://circabc.europa.eu/sd/a/3a87a7ad-858d-459e-9e45-bee034c013dd/Guidance%20Document%20No%2017%20-%20Direct%20and%20indirect%20inputs.pdf (accessed on 26 February 2022).
41. Quevauviller, P. *Groundwater Science and Policy: An International Overview*; Royal Society of Chemistry: London, UK, 2007; p. 796.
42. Brenčič, M.; Prestor, J.; Kompare, B.; Matoz, H.; Kranjc, S. Integrated approach to delineation of drinking water protection zones = Integrirani pristop k določanju vodovarstvenih območij. *Geologija* **2009**, *52*, 175–182. [CrossRef]
43. Environmental Atlas of Slovenia. Water Protection Areas. 2021. Available online: http://gis.arso.gov.si/atlasokolja/profile.aspx?id=Atlas_Okolja_AXL@Arso&culture=en-US (accessed on 12 April 2021).
44. ZRSVN—Zavod RS za Varstvo Narave (Institute of the Republic of Slovenia for Nature Conservation). *Areas of Groundwater Dependent Ecosystems (Digital Polygon Layer)*; ZRSVN: Ljubljana, Slovenia, 2014.
45. Mezga, K.; Janža, M.; Prestor, J.; Koren, K.; Šram, D. Groundwater dependent ecosystems—groundwater status indicators. *Nat. Slov.* **2016**, *18*, 35–42.
46. ARSO. Methodology for Groundwater Quality Assessment. Slovenian Environment Agency. 2009. Available online: http://www.arso.gov.gov (accessed on 12 November 2020).
47. European Commission. *Technical Report on Groundwater Dependent Terrestrial Ecosystems. WFD CIS Guidance Document No. 6. Technical Report 056-2011*; European Commission: Luxembourg, 2011; p. 32. [CrossRef]
48. Uhlman, K. The Geochemistry of Boron in a Landfill Monitoring Program. *Ground Water Monit. Remediat.* **2007**, *11*, 139–143. [CrossRef]

Article

Analysis of Temperature Influence on Precipitation of Secondary Sediments during Water Injection into an Absorptive Well

Piotr Jakubowicz *, Teresa Steliga and Katarzyna Wojtowicz

Oil & Gas Institute—National Research Institute, 31-503 Krakow, Poland
* Correspondence: piotr.jakubowicz@inig.pl; Tel.: +48-13-436-89-41

Abstract: The extraction of hydrocarbons is associated with obtaining certain amounts of water, which is heavily contaminated with a wide range of chemical compounds that negatively affect the environment. At present, practically the only method of managing extracted reservoir waters is their injection into absorbing horizons. Large changes in parameters (pH, Eh, temperature, etc.) occurring during the extraction and storage of water, as well as the contact of the injected water with reservoir water and rock, may result in the precipitation of secondary sediments. The complexity of the injected water/native water/deposit rock system and the wide range of possible interactions do not always allow for correct interpretation of the processes and their impact on near-well zone permeability. One of the factors which has a decisive influence on dissolution/precipitation is temperature change. Applying analytical data of water with low (W-1) and high (W-2) mineralization, calculations were carried out with the use of PRHEEQC software. Changes in solubility index values were determined at ambient temperature (20 °C) and reservoir temperature (94 °C). The obtained results indicate that with increasing temperature, SI changes for a given chemical compound may run in different directions and take different values, depending on the composition of the injected water. The calculations indicate the possibility of a change in the direction of the reaction from dissolution to precipitation, which may lead to clogging of the near-well zone. Simulations of the injected water's contact with minerals present in the reservoir rock were also carried out. The obtained data indicate that these minerals, in the entire studied temperature range, dissolve in the injected water, but the solubility of anhydrite and dolomite decreases with increasing temperature. If the water is saturated with minerals at low temperature, after heating in the bed, sedimentation and blockage of rock pores may occur, which means there is a reduction in the efficiency of water injection.

Keywords: water injection; absorptive well; solubility index; PHREEQC

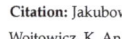

Citation: Jakubowicz, P.; Steliga, T.; Wojtowicz, K. Analysis of Temperature Influence on Precipitation of Secondary Sediments during Water Injection into an Absorptive Well. *Energies* **2022**, *15*, 9130. https://doi.org/10.3390/en15239130

Academic Editors: Robert Oleniacz and Katarzyna Grzesik

Received: 31 October 2022
Accepted: 29 November 2022
Published: 2 December 2022

Publisher's Note: MDPI stays neutral with regard to jurisdictional claims in published maps and institutional affiliations.

Copyright: © 2022 by the authors. Licensee MDPI, Basel, Switzerland. This article is an open access article distributed under the terms and conditions of the Creative Commons Attribution (CC BY) license (https://creativecommons.org/licenses/by/4.0/).

1. Introduction

Reservoir waters, often extracted in large amounts during the exploitation of crude oil and natural gas, may contain very large loads of pollutants that adversely affect living organisms [1]. There are large amounts of various substances in reservoir waters: soluble salts—mainly chlorides, but also sulphates, bicarbonates, bromides and sulfides; metal ions—sodium, potassium, calcium, magnesium, iron and manganese; heavy metals; insoluble substances—sands, clays, metal oxides and silica; remains of drilling fluids and treatment fluids; petroleum hydrocarbons—including phenols, aromatic hydrocarbons and PAHs; and chemicals dissolved in water used in order to improve its operation—corrosion inhibitors and hydrates, foaming agents, demulsifiers, etc. Treatment of such waters is difficult and costly due to strong variations in the content of individual pollutants in subsequent parts of the extracted water. The presence of a wide variety of substances makes it difficult and economically unprofitable to apply the treatment of the excavated reservoir waters to a level that enables safe discharge into surface waters or soil.

For this reason, it is preferable to remove the extracted reservoir water from the active biosphere by pumping it to absorbent horizons in hydrocarbon reservoirs. This method enables safe and inexpensive management of brines, and can be applied to intensify extraction and increase the degree of depletion of the deposit [2–4]. An alternative solution for deposit water injection is its purification and drainage to surface waters and the soil or its industrial usage, after elimination of all contaminants according to administrative law [5–7]. The diversity and huge amount of contaminants cause serious problems with water purification to the required level. Furthermore, the management of wastes which appear during water purification increases the costs of this solution.

However, injection as a preferred method of management requires proper water preparation in terms of removing components that may damage a borehole zone and limit injection efficiency.

The preparation for injection is usually carried out by removing sediment (e.g., by filtration). However, in many cases this is insufficient due to the presence of fine particles of suspended solids that are not retained on the filters, as well as iron oxide deposits formed upon contact with air.

Due to their properties, these oxides cause difficulties in filtration (clogging of filters), and may also precipitate after filtering the water. Their colloidal nature means that, together with the injected water, they can deeply penetrate the near-well zone and cause damage which is extensive and difficult to remove.

A proper solution is a water treatment installation using appropriately selected processes to remove pollutants: removal of iron and manganese ions (by air blow oxidation, absorption on the bed in ion exchange process or masking their presence with complexing compounds), coagulation (precipitation of pollutants in a form of sediment) with flocculation (agglomeration of suspended particles and colloids) or using a ballasting agent [8–10] followed by sedimentation and/or filtration to remove solid particles from the water phase [11–14]. Due to the large variation in the content of individual components in the extracted reservoir waters, it is not possible to determine optimal doses of chemicals and parameters of individual treatments at one time. It is necessary to carry out the optimization procedure for individual stages of the preparation process and to control the effectiveness individually for each batch of formation water intended for injection (Figure 1) [15].

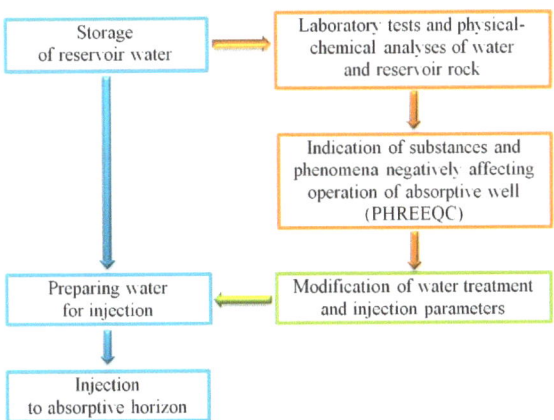

Figure 1. Scheme of water injection process with necessary laboratory and modelling research.

Membrane systems can also be used in water treatment to remove petroleum pollutants and to obtain water with low salinity. Research on the application of these methods is most often applied in areas suffering from water shortages [16].

However, even the correct and effective conduction of the process of water preparation for injection may not protect the absorptive well against progressive clogging. Undesirable chemical processes may occur when injected into a reservoir of water with a different chemical composition (the presence of incompatible components).

Precipitation of sediment particles and suspensions may take place during contact of the injected water with native water present in the reservoir and with the reservoir rock, as well as in the case of the successive injection of water batches containing components reacting with each other with precipitation. Sludge formation reactions can also be triggered by changing physical and chemical properties of the water (e.g., changing the pH, redox potential, temperature, etc.) [17,18].

Water in a deposit has its temperature (several dozen °C on average). It decreases during water excavation, separation, transport, storage and preparation for injection. The temperature of treated water is usually the same as the surrounding temperature. During injection into a well, flowing water has contact with zones of increasing temperature. Reaching an absorptive horizon, the water heats up to the deposit temperature. To a certain extent, water temperature can be controlled by the following treatments: limiting the time of the presence of water on a surface, thermal isolation of the installation, a change in injection speed (a decrease in speed results in higher temperature growth before the water reaches an absorptive horizon) or heating the water before injection. However, these methods cannot be technically and economically justified. The exception is a situation in which excavated water (after separation from hydrocarbons) can be injected directly into an absorptive horizon. No sediments or suspensions in the water, low hydrocarbon content, and a lack of contact with oxygen during preparation and injection are required.

The system of injected water/reservoir water/reservoir rock is characterized by a high degree of complexity, taking into account all factors influencing changes in physical–chemical properties and a wide range of interactions between individual components. The available analytical data do not always enable correct assessment of the direction of precipitation/dissolution and determination of their impact on the stability of the system, particularly on the permeability of the near-well zone.

Computer simulations enable assessment of direction of sediment dissolution/precipitation when water is injected into the deposit [19–22].

At present, there are several specialized hydrogeochemical modeling systems available, e.g., TOUGHREACT [22–25], ChemPlugin [26], PHREEQC [27–29], and FEDLOW (piChem module) [30,31], which can be used to solve problems related to the chemistry and reactivity of water, mass and energy transport, secondary mineral precipitation, exploitation of geothermal deposits, etc. [32–36].

PHREEQC software (pH-REdox-EQuilibrium) [37–39] was selected for tests to perform simulations on highly concentrated brines. This program is widely used to study issues related to the flow of brine through porous media and accompanying phenomena [40–43], especially in the processes of deposits flooding (enhanced oil recovery) in oil mining [44–47].

This software package calculates solubility indexes (SIs) which can be used to assess precipitating/dissolving ability [18,45,48]. Moreover, it allows the determination of the effect of temperature change on the solubility of individual minerals, which may be key information for limiting damage to the near-well zone during the process of injecting water into the absorptive horizon.

PHREEQC is software with a wide range of applications. It can be used in simulations of processes and geochemical environments. However, it has several limitations which ought to be taken into account. One of the limitations of water models is a lack of inner cohesion in implemented databases. The pitzer.dat database defines the most consistent water model, but its contents of ingredients are limited. All remaining databases include sets of constant equilibrium logarithms ($logK_s$) and enthalpies of reactions coming from the literature. There were no systematic attempts to determine which tests were performed in order to determine the assumed individual parameters of the $logK_s$. In addition, there was no determination of whether the simulation results (obtained by application of a certain model) are compatible with original experimental data. Information included in the files of databases delivered with the software should be treated as preliminary. The precise selection of solutions' contents and thermodynamic data is in the hands of the user.

The presented research results in a decrease in contaminants emitted to the natural environment during oil and natural gas excavation. Improved methods of brine management lead to significant reductions in the use of environmental resources by the oil industry.

This article describes innovative applications of PHREEQC to determine problems associated with near-well zone colmatation. Difficulties are caused by sediments which can appear in the injection of consecutive parts of the water of diversified properties into absorptive horizons. Obtaining the above-mentioned information enables modification of the parameters of the preparation and injection of waters into an absorptive horizon. In addition, it leads to a reduction in the risk of a decrease in rock permeability by sediments. It would result in less frequent application of expensive treatments such as near-well zone decolmatation.

These benefits can encourage industry representatives to perform more advanced planning of deposit water injection. This involves selecting a proper deposit to inject with a certain type of water, performing research on the compatibility of injected waters and increasing the effectiveness of water preparation. This can result in extensions of the time of failureless work of absorptive horizons as well as decreases in the costs of management of excavated waters.

2. Materials and Methods

Research material consisted of W-1 and W-2 waters from wells exploiting natural gas from reservoir in Cechsztyn limestone, with highly differentiated properties (Table 1). In the tested waters, the following parameters were determined: reaction, redox potential and sodium content (pH/mV/Ion/°C/F—ION 700 m—measurement error 2%), chemical and biological oxygen demand, total organic carbon, anionic and nonionic surfactants and potassium (Hach Lange DR 3900 photometer—measurement error 4%), density (Anton Paar 35 N—measurement error 5%), content of dissolved and undissolved substances, residue after ignition at 600 °C and content of organic substances (analytical balance WAA 220/C/2—measurement error 3%), anions (Cl^-, Br^-, CO_3^{2-}, HCO_3^-, S_2^-, SO_4^{2-}), calcium and magnesium, carbonates, bicarbonates and chlorides (complexometric, acid-base and argentometric titration—analysis error 4%), sulphates, iron, manganese and selected heavy metals (UV/VIS Lambda 35 spectrophotometer—measurement error 4%).

Table 1. Results of physical–chemical analyses of separator water samples W-1 and W-2.

Analysis	Unit	W-1	W-2
pH		5.9	4.8
Eh	g/cm^3	−108	−117.8
Density (20 °C)	mV	0.997	1.182
Total dissolved substances	mg/dm^3	551	306,428
Residue of roasted (in 600 °C)	mg/dm^3	318	288,904
Total suspended solids	mg/dm^3	76	159
COD	mg O$_2$/dm^3	15,023	13,589
BOD	mg O$_2$/dm^3	1875	2258
TOC	mg/dm^3	1004	1059
TPH	mg/dm^3	64	284
Organic substances (dichloromethane extract)	mg/dm^3	91	1102
Anionic surfactants	mg/dm^3	1.23	18.9
Nonionic surfactants	mg/dm^3	247	1.73
Chloride Cl^-	mg/dm^3	129	176,615
Sulphates SO_4^{2-}	mg/dm^3	4.3	189
Carbonates CO_3^{2-}	mg/dm^3	–	–
Bicarbonates HCO_3^-	mg/dm^3	215	169
Nitrates NO_3^-	mg/dm^3	–	–

Table 1. Cont.

Analysis	Unit	W-1	W-2
Ammonium NH_4^+	mg/dm^3	–	–
Phosphates PO_4^{3-}	mg/dm^3	–	–
Bromides Br^-	mg/dm^3	4.12	249.3
Sodium Na^+	mg/dm^3	61.9	68,841
Potassium K^+	mg/dm^3	28.6	588
Calcium Ca^{2+}	mg/dm^3	18.6	35,258
Magnesium Mg^{2+}	mg/dm^3	12.7	4974
Ferrous ion Fe^{2+}	mg/dm^3	2.10	6.50
Ferric ion Fe^{3+}	mg/dm^3	16.00	56.40
Manganese Mn^{2+}	mg/dm^3	3.91	7.05
Copper Cu	mg/dm^3	0.021	0.009
Lead Pb	mg/dm^3	0.068	0.035
Zinc Zn	mg/dm^3	0.651	0.358
Tin Sn	mg/dm^3	0.023	0.51
Nickel Ni	mg/dm^3	0.067	0.129
Cobalt Co	mg/dm^3	0.009	0.028
Cadmium Cd	mg/dm^3	0.003	0.048
Strontium Sr	mg/dm^3	0.061	3012
Barium Ba	mg/dm^3	0.038	81.0
Silicon Si	mg/dm^3	3.18	4.26
Aluminum Al	mg/dm^3	0.061	0.056

W-1 water was characterized by low mineralization at a level of 551 mg/dm^3 and suspension content of 76 mg/dm^3, while W-2 water sample was highly mineralized (306,428 mg/dm^3) and suspension content was 159 mg/dm^3.

Chemical analyses were supplemented by mineralogical analysis of core material samples from the injection well (Figure 2). The quantitative analysis of the mineral composition of the core samples was performed using the X-ray diffraction method (XRD) at the Department of Drilling Geophysics, Oil and Gas Institute—National Research Institute [49].

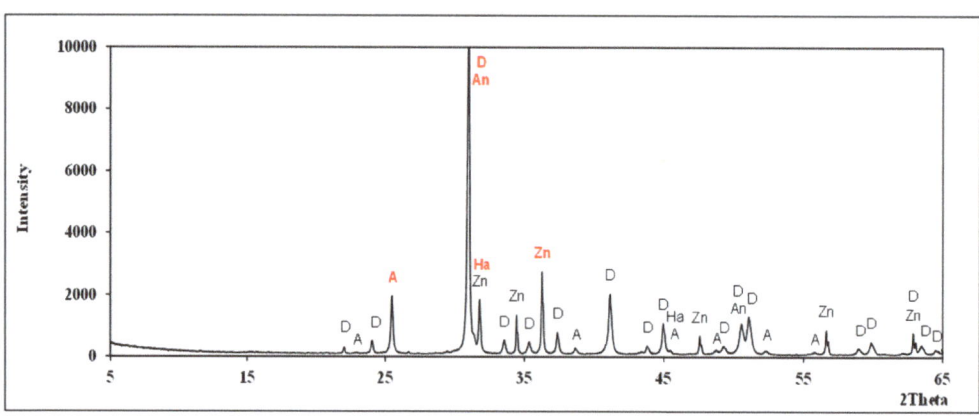

Figure 2. Results of mineral composition analysis of core material from injection well (D—dolomite, A—anhydrite, An—ankerite, Ha—halite, Zn—zincite—reference material).

Mineralogical analysis showed that the main component of reservoir rocks is dolomite (74.6%), whereas the other components are anhydrite (13.1%) and ankerite (11.3%), calcium and iron carbonate with the formula $CaFe(CO_3)_2$ and a slight addition of halite (1.0%).

The solubility index (SI) is an indicator suggesting the possible direction of reaction of dissolution/precipitation of minerals during mixing of water or in contact between water and a deposit rock. PHREEQC software was used to calculate the SI when considering injection of separator waters into the absorptive horizon.

The values of the solubility index calculated for individual components (minerals or chemical compounds) indicate whether it is possible to dissolve a specified component in water of a given composition (negative SI values), or whether their precipitation from the solution should be expected (positive SI values).

The assumption of a value close to "zero" by the calculated solubility index suggests the existence of an equilibrium state in which even small changes in water parameters (e.g., change in temperature, reaction or the content of one of the components) may cause a shift towards dissolution or precipitation of a defined substance.

There are three databases that can be used to calculate the solubility indexes: phreeqc.dat [50], pitzer.dat [51,52] and wateq4f.dat [53]. Reports in the literature show that the pitzer.dat database should be used to simulate highly concentrated brines, which is based on the calculation of ionic strength of the solution (Pitzer theory) [54] and not concentrations of components. This database, unfortunately, has a small number of components (about 40) for which it is possible to perform calculations. The other two databases should be used for calculations in solutions in the Debye–Hückel region. The phreeqc.dat database has 60 components, while the waeq4f.dat includes over 160 items.

In the vast majority of cases, the extracted formation water is prepared for injection in batches, the volume of which depends on the size of the storage tank in which it is collected. The entire process of separating water from hydrocarbons and its storage reduces temperature to ambient temperature. On the other hand, during the injection, water temperature rises to reservoir temperature. This can cause significant changes in solubility of individual minerals and affect permeability of the near-well zone.

The results of laboratory analyses of W-1 and W-2 water samples' composition were used to calculate solubility indexes for individual minerals (PHREEQC). In order to capture influence of temperature on solubility of individual components (minerals and chemical compounds), SI calculations were performed for two temperatures: 20 °C, assumed as the temperature of water prepared for injection, and 94 °C, the reservoir temperature of the injection well.

Taking into account the distinct influence of water temperature changes on the direction of the dissolution/precipitation, simulations of changes in solubility of minerals (forming reservoir rock) during contact with the injected water as a function of temperature were also carried out. The simulations assumed the contact of 1 kg of deposit rock with 1 kg of injected water. Due to the lack of ankerite (and other minerals with similar composition) in the databases of the PHREEQC software, this core component was not included in the calculations.

3. Results and Discussion

The results of the analyses show that the mere removal of sediments and suspensions from the water phase is not enough to pump the water safely into the absorptive horizon [55]. Water samples W-1 and W-2 were filtered through a filter to remove sediment and left in contact with the air for 48 h. After this time, the water was filtered again, obtaining sediments with masses of 14.35 mg/dm^3 (W-1) and 55.40 mg/dm^3 (W-2), respectively (Figure 3).

In the tested water samples, iron ions were present. These ions are relatively easily oxidized with oxygen from the air, which resulted in gradual precipitation of sediments in the filtered water. An increase in the Eh of the water can occur not only through contact with the air, but also through degassing the water from hydrocarbons (methane) dissolved in it under high pressure, as well as by mixing with water with oxidizing properties. This proves that in the case of increases in the electrochemical potential of the water, clogging of the near-well zone may occur due to sediment formation during the water's injection, and even after injection into the reservoir.

Figure 3. Effect of water aeration (water samples W-1 and W-2: a—raw water without air contact, b—after aeration, c—filter cake).

Therefore, using analytical data from the W-1 and W-2 samples' water composition, calculations of solubility indexes were carried out using PHREEQC in order to determine the possibility of precipitation of sediments from the water during injection into the absorptive horizon. This program is widely used to simulate the course of reactions and processes occurring in a porous medium such as a hydrocarbon deposit (Table 2).

Table 2. Major findings from the literature (last 2 years) concerning phenomena accompanying the injection of water.

No.	Main Issues	Reference
1	Interactions between brine and rock minerals in static and dynamic system.	[40]
2	Study of hydrochemical simulations of a dual-layer geothermal reservoir to the long-term impact of barite scale formation on well injectivity.	[41]
3	Description of mechanistic model constructed for low-salinity water injection to consider geochemical reaction issues in low-salinity flooding among surface sites and aqueous solution.	[18]
4	Integrated open-source simulator to model hydrogeochemical processes at various scales of interest including pore-scale and reservoir-scale.	[42]
5	Modeling (with PHREEQC software) of mineral precipitation and deposition in the porous media controlled by deep bed filtration model.	[44]
6	Study of fine particle migration in the rock causing formation damage and permeability impairment.	[43]
7	Investigation of the carbonate/brine interactions, using geochemical modeling, during low-salinity water injection for enhanced oil recovery (EOR).	[45]
8	Modeling of different geochemical effects such as multivalent cation exchange and mineral dissolution during flow and transport in low-salinity waterflooding.	[46]
9	Comparison of thermodynamic data files from PHREEQC software package and influence of TDF choice on modeling results.	[47]
10	Studies of influence of anhydrite on wettability of calcite rock during low-salinity water flooding.	[56]

3.1. Temperature Impact on Water Injection Process

Temperature change is a very important parameter that influences the processes of dissolution/precipitation of minerals accompanying injection of water into the reservoir. Calculations of the solubility indexes (SIs) were performed for two temperatures, 20 °C and 94 °C, with the use of all three databases, in order to maximize the obtained information on the behavior of minerals and chemical compounds. The calculation results for the water intended for injection are shown in the figures below (Figures 4–6).

Figure 4. Comparison of SI calculated for W-1 and W-2 water samples using phreeqc.dat database at 20 °C (ambient) and 94 °C (deposit) temperatures.

Considering the values of the solubility index (SI) calculated with the application of the phreeqc.dat database (calculations by ionic strength of solutions, taking into account high salt concentrations), it can be seen that the dissolution/precipitation potentials of individual minerals are very different (Figure 4). For most substances, the equilibrium in the tested water solutions is shifted in the direction of dissolution. In the direction of precipitation, the balance is shifted mainly for iron-containing minerals (goethite, hematite or pyrite) and for aluminosilicates (montmorillonite, mica or kaolinite). In the case of manganese, which is usually indicated with iron, as a product easily precipitating from the solution in the form of oxide and hydroxide deposits, the calculations showed high negative SI values. This indicates an equilibrium shifted strongly towards dissolution of manganese-containing compounds and no risk of precipitation.

Figure 5. Comparison of SI calculated for W-1 and W-2 water samples using pitzer.dat database at 20 °C (ambient) and 94 °C (deposit) temperatures.

On the other hand, when considering the impact of temperature change on the ability of the injected water to dissolve individual minerals, significant differences should be noted, both in the size and the direction of changes in the calculated solubility indexes. In individual waters, the values of the SI calculated for the same minerals/chemical compounds may not only change their value, but also assume different directions of changes. For example, in the case of FeS, a significant decrease in solubility (IS from −76.75 to −19.98) was noted for the W-1 sample, while for the W-2 sample there was an increase in solubility (IS from +10.86 to −0.66). The increase in temperature may also change the direction of the dissolution/precipitation. In the case of alunite for the W-1 water, a change in the index indication from a slight tendency to precipitation to a fairly strong tendency to dissolve was observed, while for the W-2 stream, a change in the SI value from clearly negative (−17.12), i.e., from dissolving at 20 °C to precipitate alunite (SI = +1.68) at 94 °C was observed. Such dependencies are practically impossible to capture without computer simulations, and may in some cases have a decisive impact on the course and efficiency of water injection.

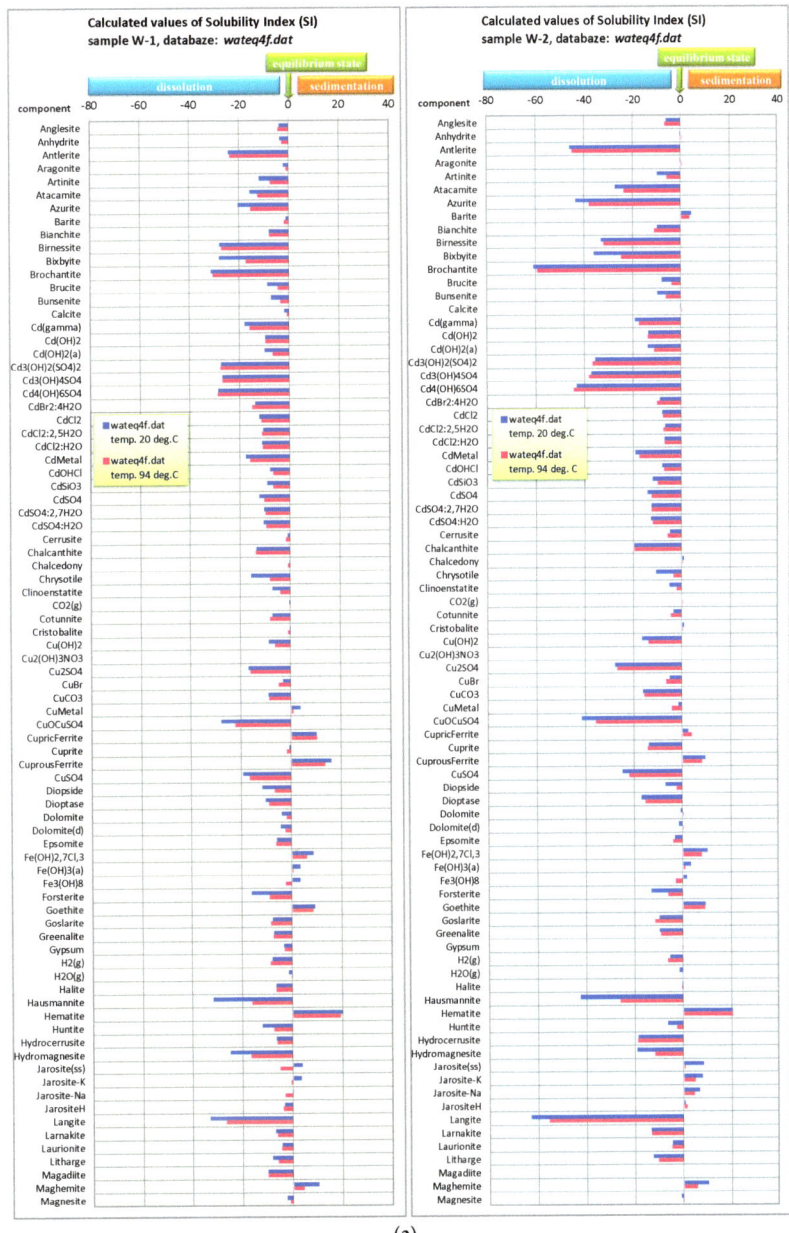

(a)

Figure 6. Cont.

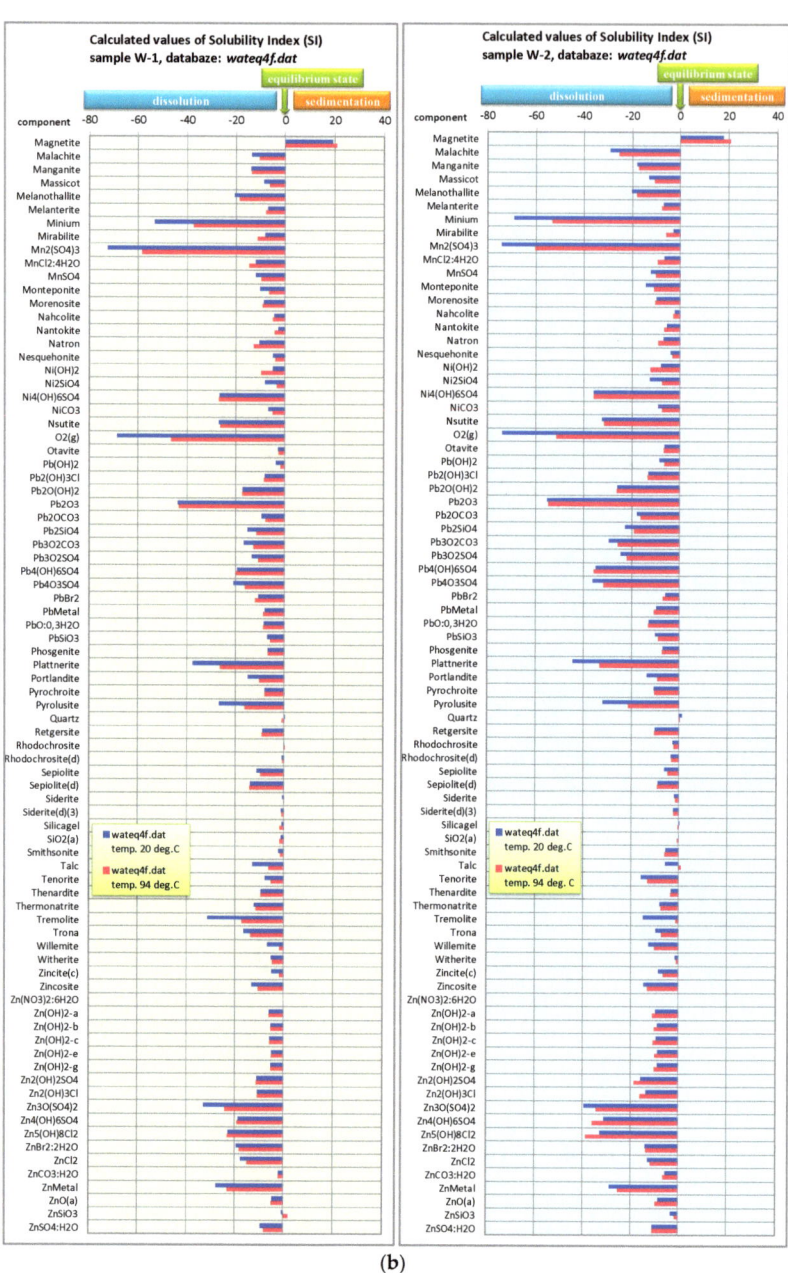

Figure 6. (**a**) Comparison of SI calculated for W-1 and W-2 water samples using wateq4f.dat database at 20 °C (ambient) and 94 °C (deposit) temperatures. (**b**) Comparison of SI calculated for W-1 and W-2 water samples using wateq4f.dat database at 20 °C (ambient) and 94 °C (deposit) temperatures.

Similar calculations for the tested water samples intended for injection into the deposit were made with the use of pitzer.dat (Figure 5).

There is a relatively small number of components in this database and due to the lack of iron-containing minerals, its application in modeling the processes is significantly limited. Considering the obtained results, it should be stated that for both W-1 water with low mineralization and W-2 with high mineralization, almost all components included in the calculations are in a state of unsaturation in the solution and their equilibrium is shifted towards dissolution. Usually, large differences in mineralization and the content of individual dissolved substances in water do not cause a greater variation in the calculated mineral solubility indexes. Only in the case of a few substances (anhydrite, aragonite, calcite, celestite, dolomite and carbon dioxide) was there a slight shift of the equilibrium towards precipitation (or gas phase separation for CO_2).

A significant change in temperature from 20 °C to 94 °C in the performed simulations usually did not cause significant changes in the values of the solubility indexes. In the case of the W-2 water sample only, for the components showing a slight tendency to precipitate at 20 °C, a slight increase in the SI value (enhancement of the tendency to precipitate) was noted for the temperature of 94 °C.

Considering the results from the wateq4f.dat database, it was noted that water–mineral balance, for both W-1 (low-mineralization) and W-2 (high-mineralization) water, is usually shifted towards dissolving the minerals. The calculated solubility indexes indicated that precipitation of iron-containing minerals (hematite, goethite, magnetite and other iron compounds) should be expected for both analyzed samples of water, as well as, to a small extent, barite and quartz for W-2 high-mineralized water (Figure 6a,b).

An increase in water temperature from 20 °C to 94 °C, for most of the substances in the dissolution area, indicated a decrease in solubility. For only some compounds was there a slight increase in solubility, while for substances in the area of precipitation, the calculated SI values showed a decrease in the tendency to precipitate.

In the cases of only two iron-containing components, $Fe_3(OH)_8$ for both tested waters and jarosite for W-1 water, a change in the course of the reaction was found from precipitation at a temperature of 20 °C to dissolution at a temperature of 94 °C.

Attention should also be paid to a strong tendency to dissolve (absorb) oxygen in the tested waters, indicated in the phreeqc.dat and wateq4f.dat databases. In the case of contact with the air, the water is oxygenated quickly, which results in the precipitation of iron oxides and hydroxides.

The algorithm used in the wateq4f.dat database does not enable taking into account the influence of higher ion concentrations on the calculated solubility indexes. Nevertheless, the presence of a much larger number of chemical compounds in this database may lead to obtaining useful information. The calculation results from wateq4f.dat are generally more similar to the pitzer.dat database (despite the existing differences in the calculated SI values) than phreeqc.dat; however, due to the small number of common minerals found in all databases, it is difficult to determine existing similarities and differences.

3.2. Temperature Influence on Solubility of Rock Matrix Materials in Injected Water

Considering the influence of temperature changes on the solubility of minerals that build the reservoir rock of the absorbent horizon in contact with the W-1 water (Figure 7), it should be noted that both anhydrite and dolomite obtain maximum solubility at low temperatures (20 °C) at levels of 4.66 g/kg and 0.45 g/kg of water, respectively. The amount of halite dissolved from the deposit rock for this system is constant at 9.93 g/kg of water, regardless of temperature.

With the temperature increase, dolomite solubility decreased to the level of 0.23 g/kg (48.89%) of water for temperature of 94 °C. On the other hand, for anhydrite up to a temperature of about 30 °C, a slight decrease in solubility was noted, and at higher temperatures, there was an increase in the slope of the curve and a greater decrease in the solubility of the mineral to the level of 1.77 g/kg of water (reduction by 62.02% by weight).

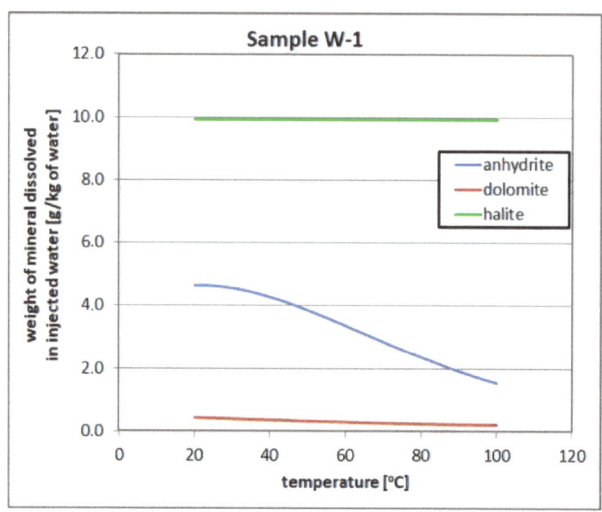

Figure 7. Changes in solubility of rock matrix minerals in W-1 injected water as function of temperature.

Assuming that chemical equilibrium is established, i.e., reaching the mineral saturation point for a given temperature, heating the injected water results in lower solubility for anhydrite and dolomite and a gradual precipitation of these minerals. For a continuous flow of water processes, this involves the leaching of minerals near the injection point and their precipitation at an increasing distance as a phenomenon of heating of the injected water.

On one hand, this seems to be beneficial, because the water washes away the minerals which build the reservoir rock and reduces the effects of clogging of the near-well zone by sediments injected with the water. On the other hand, in a longer period of time and assuming cyclical injection of water into the deposit, one can expect the rock pores to overgrow at a certain distance from the borehole zone by precipitating anhydrite and dolomite deposits, which may result in a reduction of permeability.

A simulation carried out for the W-2 sample shows a relatively low solubility of minerals forming the deposit rock in the injected water with a high content of dissolved salts (Figure 8).

As in the case of the W-1 water sample presented above, for the W-2 sample, the solubility of dolomite slightly decreases from 0.18 g/kg of water at 20 °C to 0.13 g/kg at 94 °C. In the case of anhydrite, a slightly higher solubility of the mineral at low temperatures was observed (maximum solubility at 28 °C is 0.491 g/kg of water), which decreases to 0.15 g/kg of water when heating the water to the deposit temperature (this is similar to the level calculated for dolomite). At temperatures above 94 °C, the solubility of anhydrite continues to decrease, reaching values below dolomite solubility.

Simulations of the contact of tested waters (intended for injection) with the deposit rock indicate different responses of the systems. For both water samples injected into the deposit, contact with the deposit rock dissolves the minerals that build the rock matrix, halite in particular. In the case of water sample W-2 with high mineralization, the mass of minerals that can be dissolved is much lower than in the case of water with low mineralization. All tested minerals are found throughout the temperature range in the dissolution area. However, it should be noted that the solubility of minerals decreases significantly with increasing temperature. For W-2 water, the calculated maximum dolomite solubility decreased 2.5 times, whereas for anhydrite it decreased 9.5 times. If the reservoir water is saturated with a given mineral at a low temperature, its increase may result in sedimentation and blocking of rock pores. In the case of halite, its solubility increases all the time with increasing temperature.

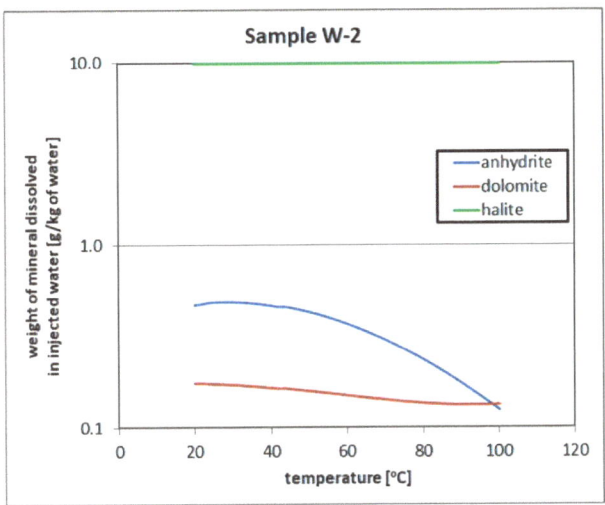

Figure 8. Changes in solubility of rock matrix minerals in W-2 injected water as function of temperature.

Obtaining information on possible undesirable processes taking place in deposit water injection into a certain horizon could lead to a decision about water injection into a different reservoir or have an impact on more advanced preparations of water for injection by the elimination of untypical contents. The elimination of problems related to decreases in near-well zone permeability by developed sediments will result in avoidance or reduction in frequency of expensive and not always effective treatments of rinsing and/or acidification of a well. As a result, there will be a significant decrease in the costs of deposit water management.

One of the important issues related to the production of hydrocarbons is the carbon footprint, which shows the impact of this human activity on the environment. At present, there is no unitary methodology which could determine a carbon trace and the entire life cycle of a product for the energetics sector, including for hydrocarbon extraction. Several main phases can be observed in the life cycle of natural gas and oil. These are searching for deposits, extraction, processing (treatment/purification), transport, storage, distribution and usage [57–61].

The management of extracted waters through their injection into a deposit enables a decrease in carbon traces. It is an effect of the reduction of the number of pollutants (including hydrocarbons) emitted to the active biosphere. Moreover, low usage of materials and energy in water injection (in comparison to complete purification of brine and waste management) should be taken into consideration. The proper preparation of water for injection leads to a reduction in carbon traces according to the long-term failure-free work of a borehole. This process results in a decrease in the frequency of decolmatation of a near-well zone, which means there is usage of a significant amount of equipment as well as chemicals in order to prepare treatment liquids, and the necessity of after-treatment liquid management.

The research, owing to the possibility of a reduction in the costs of brine management as well as its possible usage in order to increase a level of deposit exploitation (flooding), could result in mitigation of the energy crisis.

4. Conclusions

Considering the obtained values of the solubility indexes (SI) calculated with the application of databases available in PHREEQC software, it can be seen that the dissolution/precipitation potential of individual minerals is different. For most substances, the equilibrium in the injected water solutions is shifted in the direction of dissolution. In the direction of precipitation, the balance is shifted mainly for iron-containing minerals (goethite, hematite and pyrite) and for aluminosilicates (montmorillonite, mica and kaolinite). On the other hand, manganese, usually considered together with iron as a product easily precipitating from the solution in the form of oxide and hydroxide deposits, shows equilibrium shifted strongly towards the dissolution of minerals (high negative SI values).

The increase in temperature to the temperature of the deposit causes different reactions in individual minerals and chemical compounds. Depending on the chemical composition and physical–chemical parameters of a water sample, changes in the solubility of a specific mineral may take different directions and take various values, including a transition between the areas of precipitation and dissolution.

When interpreting the results of simulations, particular attention should be paid to individual components in the sensitive zone close to the equilibrium state (SI = 0). In this area, even slight changes in the properties of the water can change the solubility and formation of sediments that can damage the permeability of the reservoir rocks.

The conducted simulations of the contact of water with the deposit rock indicate the different activities of individual systems. For the tested waters, contact with the deposit rock dissolves the minerals which build the rock matrix, especially halite. However, it should be noted that the solubility of dolomite and anhydrite decreases significantly with increasing temperature. If the reservoir water is saturated with a given mineral at a low temperature, its heating may result in sediment precipitation. In the case of halite, solubility increases all the time with increasing temperature.

The saturation of water with halite may also result in changes in the solubility of anhydrite and dolomite. A decrease in the amount of dissolving minerals is clearly visible along with an increase in the overall mineralization of the water. For brine with a high content of chlorides (W-2), the maximum solubility of dolomite decreased by 2.5 times, and the solubility of anhydrite decreased by 9.5 times compared with water with low mineralization (W-1). The changes in solubility observed in the simulations proceed in various directions, strongly dependent on the composition and physical and chemical parameters of water, and are difficult to predict without performing calculations.

Author Contributions: Conceptualization, P.J., T.S. and K.W.; methodology, P.J., T.S. and K.W.; validation, P.J., T.S. and K.W.; formal analysis and investigation, P.J., T.S. and K.W.; resources, P.J., T.S. and K.W.; writing—original draft preparation, P.J., T.S. and K.W.; writing—review and editing, P.J., T.S. and K.W.; visualization, P.J., T.S. and K.W.; supervision, P.J. All authors have read and agreed to the published version of the manuscript.

Funding: This research was financially supported by Polish Ministry of Science and Higher Education within statutory funding for Oil and Gas Institute-National Research Institute.

Data Availability Statement: All supporting data have been included in this study and are available from the corresponding authors upon request.

Conflicts of Interest: The authors declare no conflict of interest.

References

1. Steliga, T.; Jakubowicz, P.; Kapusta, P. Changes in toxicity during treatment of wastewater from oil plant contaminated with petroleum hydrocarbons. *JCTB* **2015**, *90*, 1408–1418. [CrossRef]
2. Muggeridge, A.; Cockin, A.; Webb, K.; Frampton, H.; Collins, I.; Moulds, T.; Salino, P. Recovery rates, enhanced oil recovery and technological limits. *Phil. Trans. R. Soc. A* **2014**, *372*, 1–25. [CrossRef] [PubMed]
3. Rubinstein, J.L.; Mahani, A.B. Myths and Facts on Wastewater Injection, Hydraulic Fracturing, Enhanced Oil Recovery and Induced Seismicity. *Seismol. Res. Lett.* **2015**, *86*, 1060–1067. [CrossRef]
4. Lubaś, J.; Stopa, J.; Warnecki, M.; Wojnicki, M. Możliwości zastosowania zaawansowanych metod wspomagania wydobycia ropy naftowej ze złóż dojrzałych. *Nafta-Gaz* **2019**, *1*, 24–28. [CrossRef]
5. Purnima, M.; Paul, T.; Pakshirajan, K.; Pugazhenthi, G. Onshore oilfield produced water treatment by hybrid microfiltration-biological process using kaolin based ceramic membrane and oleaginous *Rhodococcus opacus*. *Chem. Eng. J.* **2023**, *453*, 139850. [CrossRef]

6. Weschenfelder, S.E.; Louvisse, A.M.T.; Borges, C.P.; Meabe, E.; Izquierdo, J.; Campos, J.C. Evaluation of ceramic membranes for oilfield produced water treatment aiming reinjection in offshore units. *J. Pet. Sci. Eng.* **2015**, *131*, 51–57. [CrossRef]
7. Fakhru'l-Razi, A.; Pendashteh, A.; Abdullah, L.C.; Biak, D.R.A.; Madaeni, S.S.; Abidin, Z.Z. Review of technologies for oil and gas produced water treatment. *J. Hazard. Mater.* **2009**, *170*, 530–551. [CrossRef]
8. Janocha, A.; Kluk, D. Research on the possibility of using loading materials to support the treatment of reservoir waters. *Nafta-Gaz* **2021**, *4*, 255–263. [CrossRef]
9. Lapointe, M.; Barbeau, B. Characterization of ballasted flocs in water treatment using microscopy. *Water Res.* **2016**, *90*, 119–127. [CrossRef]
10. Zafisah, N.S.; Ang, W.L.; Mohammad, A.W.; Hilal, N.; Johnson, D.J. Interaction between ballasting agent and flocs in ballasted flocculation for the removal of suspended solids in water. *J. Water Process Eng.* **2020**, *33*, 101028. [CrossRef]
11. Jakubowicz, P. Wybrane problemy zagospodarowania odpadowych wód kopalnianych. *Nafta-Gaz* **2010**, *5*, 383–389.
12. The, C.Y.; Budiman, P.M.; Shak, K.P.Y.; Wu, T.Y. Recent Advancement of Coagulation–Flocculation and Its Application in Wastewater Treatment. *Ind. Eng. Chem. Res.* **2016**, *55*, 4363–4389. [CrossRef]
13. Jakubowicz, P.; Steliga, T. Efektywność działania nowoczesnych koagulantów glinowych w warunkach obniżonego pH wód z formacji łupkowych. *Nafta-Gaz* **2017**, *3*, 169–176. [CrossRef]
14. He, W.; Xie, Z.; Lu, W.; Huang, M.; Ma, J. Comparative analysis on floc growth behaviors during ballasted flocculation by using aluminum sulphate (AS) and polyaluminum chloride (PACl) as coagulants. *Sep. Purif. Technol.* **2019**, *213*, 176–185. [CrossRef]
15. Jakubowicz, P.; Steliga, T. Assessment of the Main Threats to Injection Well Damage Caused by Reservoir Waters Using Aquachem Software as Well as Laboratory Tests Application. *Nafta-Gaz* **2012**, *10*, 655–660.
16. Igunnu, E.T.; Chen, G.Z. Produced water treatment technologies. *Int. J. Low-Carbon Technol.* **2014**, *9*, 157–177. [CrossRef]
17. Kluk, D. Badania procesu mieszania wód zatłaczanych z wodami złożowymi o zróżnicowanych potencjałach elektrochemicznych. *Nafta-Gaz* **2011**, *2*, 98–106.
18. Abbasi, P.; Abbasi, S.; Moghadasi, J. Experimental investigation of mixed-salt precipitation during smart water injection in the carbonate formation. *J. Mol. Liq.* **2020**, *299*, 112131. [CrossRef]
19. Dobrzyński, D. Modelowanie geochemiczne narzędziem poznania geochemii systemów wód podziemnych. Przykłady zastosowań, aktualny stan w Polsce. *Przegląd Geol.* **2006**, *54*, 976–981.
20. Li, X.; He, X.; Yang, G.; Zhao, L.; Chen, S.; Wang, C.; Chen, J.; Yang, M. Study of groundwater using visual MODFLOW in the Manas River Basin, China. *Water Policy* **2016**, *18*, 1139–1154. [CrossRef]
21. Wang, G.; Wu, Q.; Yan, Z.; Zhao, N.; Duan, C.; Cheng, X.; Wang, H. Fine Prediction for Mine Water Inflow on Basis of Visual Modflow. *OGCE* **2019**, *7*, 52–59.
22. Li, L.; Wu, Y.; Chong, S.; Wen, Q. The application of TOUGHREACT in the field of energy and environment. *IOP Conf. Ser. Earth Environ. Sci.* **2020**, *569*, 012093. [CrossRef]
23. Pruess, K.; Oldenburg, C.; Moridis, G. *TOUGH2 User's Guide, Version 2.0*. LBL-43134; Lawrence Berkeley Laboratory Report: Berkeley, CA, USA, 1999.
24. Wanner, C.; Eichinger, F.; Jahrfeld, T.; Diamond, L.W. Unraveling the formation of large amounts of calcite scaling in geothermal wells in the Bavarian Molasse Basin: A reactive transport modeling approach. *Procedia Earth Planet. Sci.* **2017**, *17*, 344–347. [CrossRef]
25. Xu, T.; Sonnenthal, E.; Spycher, N.; Pruess, K. *TOUGHREACT User's Guide: A Simulation Program for Nonisothermal Multiphase Reactive Geochemical Transport in Variably Saturated Geologic Media*; Lawrence Berkeley National Laboratory: Berkeley, CA, USA, 2004.
26. Roberts-Ashly, T.L.; Berger, P.M.; Cunningham, J.A. Modeling geologic sequestration of carbon dioxide in a deep saline carbonate reservoir with TOUGH2–ChemPlugin, a new tool for reactive transport modelling. *Environ. Geosci.* **2020**, *27*, 103–116. [CrossRef]
27. Parkhurst, D.L.; Appelo, C.A.J. *User's Guide to PHREEQC (Version 2)—A Computer Program for Speciation, Batch-Reaction, One-Dimensional Transport, and Inverse Geochemical Calculations*; US Geological Survey Water-Resources Investigations: New Cumberland, PA, USA, 1999; Volume 312, pp. 99–4259.
28. Abdelaziz, R.; Merkel, B.J.; Zambrano-Bigiarini, M.; Nair, S. Particle swarm optimization for the estimation of surface complexation constants with the geochemical model PHREEQC-3.1.2. *Geosci. Model Dev.* **2019**, *12*, 167–177. [CrossRef]
29. Huber, P.; Neyret, C.; Fourest, E. Implementation of the anaerobic digestion model (ADM1)in the PHREEQC chemistry engine. *Water Sci. Technol.* **2017**, *76*, 1090–1103. [CrossRef]
30. Ren, J.M.; Yang, Y.; Hu, X.W. Application of GIS and FEFLOW in Forecasting Groundwater Flow Field of Minqin Basin. *Adv. Mater. Res.* **2011**, *368–373*, 2128–2131. [CrossRef]
31. Gao, Y.; Pu, S.; Zheng, C.; Yi, S. An improved method for the calculation of unsaturated–saturated water flow by coupling the FEM and FDM. *Sci. Rep.* **2019**, *9*, 14995. [CrossRef]
32. Lewkiewicz-Małysa, A.; Winid, B. Geologiczne i geochemiczne aspekty chłonności otworów wykorzystywanych do zatłaczania wód złożowych. *Srod.-Pomor. Tow. Nauk. Ochr. Sr. Rocz. Ochr. Sr.* **2011**, *13*, 1985–1999.
33. Hu, Y.; Mackay, E. Modelling of geochemical reactions during smart water injection in carbonate reservoirs. *Eur. Assoc. Geosci. Eng.* **2016**, *2016*, 1–5. [CrossRef]
34. Klunk, M.A.; Damiani, L.H.; Feller, G.; Conceição, R.V.; Abel, M.; De Ros, L.F. Geochemical modeling of diagenetic reactions in Snorre Field reservoir sandstones: A comparative study of computer codes. *Braz. J. Geol.* **2015**, *45*, 29–40. [CrossRef]

35. Jakóbczyk, S.; Kowalczyk, A. Zastosowanie modelowania geochemicznego do oceny warunków kształtowania się składu chemicznego wód podziemnych w rejonie ujęcia Gliwice Łabędy. *Biul. Państwowego Inst. Geol.* **2011**, *445*, 217–226.
36. Daneshgar, S.; Buttafava, A.; Callegari, A.; Capodaglio, A.G. Simulations and Laboratory Tests for Assessing Phosphorus Recovery Efficiency from Sewage Sludge. *Resources* **2018**, *7*, 54. [CrossRef]
37. Lassin, A.; Andre, L.; Lach, A. Considerations about the building of a thermodynamic database for the chemical description of highly saline systems. *Procedia Earth Planet. Sci.* **2017**, *17*, 304–307. [CrossRef]
38. Heredia, D.J. Improvement of the Numerical Capacities of Simulation Tools for Reactive Trans-Port Modelling in Porous Media. Ph.D. Thesis, Université Rennes 1, Rennes, France, 2017.
39. Uliasz-Misiak, B.; Chruszcz-Lipska, K. Aspekty hydrogeochemiczne związane z mieszaniem wód złożowych zatłaczanych do złoża węglowodorów. *Gospod. Surowcami Miner.* **2017**, *33*, 69–80. [CrossRef]
40. Zou, Y.; Zheng, C.; Sheikhi, S. Role of ion exchange in the brine-rock interaction systems: A detailed geochemical modeling study. *Chem. Geol.* **2021**, *559*, 119992. [CrossRef]
41. Tranter, M.; De Lucia, M.; Kühn, M. Barite Scaling Potential Modelled for Fractured-Porous Geothermal Reservoirs. *Minerals* **2021**, *11*, 1198. [CrossRef]
42. Soulaine, C.; Pavuluri, S.; Claret, F.; Tournassat, C. porousMedia4Foam: Multi-scale open-source platform for hydro-geochemical simulations with OpenFOAM®. *Environ. Model. Softw.* **2021**, *145*, 105199. [CrossRef]
43. Bagrezaie, M.A.; Dabir, B.; Rashidi, F. A novel approach for pore-scale study of fines migration mechanism in porous media. *J. Pet. Sci. Eng.* **2022**, *216*, 110761. [CrossRef]
44. Zhao, R.; Liu, X.; Hu, A.; Fan, C.; Shabani, A. A novel approach for modeling permeability decline due to mineral scaling: Coupling geochemistry and deep bed filtration theory. *J. Pet. Sci. Eng.* **2021**, *205*, 108995. [CrossRef]
45. Kalantariasl, A.; Tale, F.; Parsaei, R.; Keshavarz, A.; Jahanbakhsh, A.; Maroto-Valer, M.M.; Mosallanezhad, A. Optimum salinity/composition for low salinity water injection incarbonate rocks: A geochemical modelling approach. *J. Mol. Liq.* **2022**, *362*, 119754. [CrossRef]
46. Chen, Q.; Abu-Al-Saud, M.O.; Ayirala, S.C.; AlYousef, A.A. Propagation of mineral dissolution waves driven by cation exchange in low salinity waterflooding. *Fuel* **2022**, *328*, 125350. [CrossRef]
47. Lu, P.; Zhang, G.; Apps, J.; Zhu, C. Comparison of thermodynamic data files for PHREEQC. *Earth-Sci. Rev.* **2022**, *225*, 103888. [CrossRef]
48. Doubra, P.; Kamran-Pirzaman, A.; Mohammadi, A.H.; Hassanalizadeh, R. Thermodynamic modelling of scale (Calcite, Barite, Anhydrite and Gypsum) deposition from brine. *J. Mol. Liq.* **2017**, *230*, 96–103. [CrossRef]
49. Skupio, R.; Kubik, B.; Drabik, K.; Przelaskowska, A. Comprehensive interpretation of the borehole profile including anhydrite, carbonate and mudstone rocks based on non-destructive core tests. *Nafta-Gaz* **2022**, *9*, 641–653. [CrossRef]
50. Parkhurst, D.L.; Thorstenson, D.C.; Plummer, L.N. *PHREEQE—A Computer Program for Geochemical Calculations*; US Geological Survey Water-Resources Investigations: New Cumberland, PA, USA, 1980; Volume 195, pp. 80–96.
51. Pitzer, K.S. Thermodynamics of electrolytes—1. Theoretical basis and general equations. *J. Phys. Chem.* **1973**, *77*, 268–277. [CrossRef]
52. Plummer, L.N.; Parkhurst, D.L.; Fleming, G.W.; Dunkle, S.A. *A Computer Program Incorporating Pitzer's Equations for Calculation of Geochemical Reactions in Brines*; US Geological Survey Water-Resources Investigations: New Cumberland, PA, USA, 1988; Volume 310, pp. 88–4153.
53. Ball, J.W.; Nordstrom, D.K. *WATEQ4F—User's Manual with Revised Thermodynamic Data Base and Test Cases for Calculating Speciation of Major, Trace and Redox Elements in Natural Waters*; U.S. Geological Survey Open-File Report: Washington, DC, USA, 1991; Volume 185, pp. 90–129.
54. Pitzer, K.S. Theory—Ion interaction apoproach. In *Activity Coefficients in Electrolyte Solutions*; Pytkowicz, R.M., Ed.; Boca Raton: Florida, CA, USA; CRC Press: Florida, CA, USA, 1979; pp. 157–208.
55. Jakubowicz, P.; Steliga, T.; Kluk, D.; Wojtowicz, K. The possibility of application of computer simulation to predict the direction of the reaction during injection of reservoir waters. *Nafta-Gaz* **2021**, *4*, 244–254. [CrossRef]
56. Mogharrab, J.M.; Ayatollahi, S.; Pishvaie, M.R. Experimental study and surface complexation modeling ofnon-monotonic wettability behavior due to change in brine salinity/composition: Insight into anhydrite impurity in carbonates. *J. Mol. Liq.* **2022**, *365*, 120117. [CrossRef]
57. Papadopoulo, M.; Kaddouh, S.; Pacitto, P.; Prieur Vernat, A. *Life Cycle Assessment of the European Natural Gas Chain Focused on Three Environmental Impact Indicators*; Final Report; Marcogaz: Brussels, Belgium, 2011; pp. 1–189.
58. Prieur-Vernat, A.; Yoshida, S. *Opportunities and Challenges of LCA Applied to the Natural Gas Industry*; Program Committee A: Sustainability, 2012–2015 Triennium Work Report; International Gas Union (IGU): Barcelona, Spain, 2015; pp. 1–34.
59. Rogowska, D. GHG emission in the life cycle of motor fuel. Part I—guidelines for the determination of production mass balance. *Nafta-Gaz* **2014**, *9*, 639–646.
60. Sapkota, K.; Oni, A.O.; Kumar, A. Techno-economic and life cycle assessments of the natural gas supply chain from production sites in Canada to north and southwest. *J. Nat. Gas Sci. Eng.* **2018**, *52*, 401–409. [CrossRef]
61. Niemczewska, J.; Zaleska-Bartosz, J. Introduction to the evaluation of the environmental footprint for the oil and natural gas production sector. *Nafta-Gaz* **2020**, *8*, 527–532. [CrossRef]

Article

The Biological Drying of Municipal Waste in an Industrial Reactor—A Case Study

Jolanta Latosińska *, Maria Żygadło and Marlena Dębicka

Faculty of Environmental, Geomatic and Energy Engineering, Kielce University of Technology,
Al. Tysiąclecia PP 7, 25-314 Kielce, Poland; zygadlo@tu.kielce.pl (M.Ż.); marlena.debicka@gmail.com (M.D.)
* Correspondence: jlatosin@tu.kielce.pl; Tel.: +48-41-342-4571

Abstract: One of the methods of municipal solid waste (MSW) treatment is biodrying. The literature describes mainly the results obtained in a laboratory- and a pilot-scale reactor. The manuscript presents the results of MSW treatment in a full-scale bio-drying reactor (150 m^3). The reactor is operated in one of the Polish installations specializing in mechanical-biological treatment (MBT). During the 14 day period of biodrying in the reactor, the parameters of MSW such as the moisture, temperature, loss on ignition (LOI), and net heating value (NHV) were examined. The temperature of the air in the reactor was also examined. The research also included changes in the above-mentioned parameters of MSW located in three parts of the reactor: the front, middle, and back. The test results showed that the moisture content of the waste decreased from the initial level of 55% to the level of 30%. This was accompanied by an increase in the NHV from 6.3 MJ kg^{-1} to 9.6 MJ kg^{-1}. At the same time, the LOI decreased from 68% d.m. to 45% d.m. The LOI decrease is not favorable from the point of view of using MSW as refuse-derived fuel (RDF), as was expected in the final usage stage. The results have application value as the plant operator, having at their disposal the controlling of the reactor's ventilation and the temperature inside the reactor, should select the speed of the moisture removal from MSW at such a level as to minimize the LOI decrease.

Keywords: municipal solid waste; mechanical-biological treatment; biodrying; moisture controlling; net heating value controlling; RDF/SRF

1. Introduction

In order to minimize the presence of biodegradable waste in landfills, according to the European Directive [1,2] and the Polish regulations [3,4], mechanical-biological treatment (MBT) plants have been installed. The MBT plants combine the mechanical separation of different fractions present in household waste with the stabilization of the organic matter by means of biological processes. One of the main types of the MBT technology is biodrying, which, in the first stage, directs the waste into the reactor for drying prior to the extraction of the larger fraction intended for the production of RDF [5].

The mechanism of biodrying is a variation of the aerobic decomposition used within the MBT to stabilize the waste, which makes it analogous to composting, but realized in the short term [6,7]. During the process of biodrying, self-heating occurs due to the exothermic reactions of the chemical and biological processes during the transformation of the organic matter [8,9]. During the process, the temperature may rise to even 70 °C [10]. The water removal is supported by both the heat produced by the microorganisms in the biodegradation processes and the air ventilation system. The biodrying process is considered a good solution for a quick reduction in the water content of the MSW [11–14]. The method has the versatile possibilities of using the product of the treatment because it allows one to achieve the required moisture reduction, the volume reduction, and the bulk density enhancement through the effective utilization of the biological heat [11,15,16]. The method is also used for the production of high-quality solid recovered fuel (SRF)—free from hazardous substances and high in biomass content [7].

The degree of the waste drying can be varied by the correct regulation of the process parameters in the reactor [10]. The most important are the airflow rate, temperature, and moisture content. The optimum temperature of the biodrying process is 45 °C; however, even higher temperatures of the mass can be achieved in biodrying [17].

The degree of waste drying can be varied by the correct regulation of the process parameters in the reactor [10], which is applicable when the purpose of the process is to obtain a substrate for the production of an alternative fuel.

The effectiveness of the waste biodrying measured by the moisture decline is about 20–25% [11,16]. The partial decomposition of the organic matter during the biodrying should be under control when the goal is to maintain a high content of organic carbon in order to ensure a high calorific value of the waste, when the aim of the biodrying waste is to produce the RDF. After 14 days of the biodrying process in the reactor, the net heating value (NHV) of the waste could be increased by 20–40% compared to the untreated waste [10].

Most often, researchers describe the results obtained in a laboratory-scale reactor [18–22]. There are only few publications that have presented the results of research on the pilot scale [16,19]. In the literature, there are even fewer articles on bio-drying on an industrial and full scale [13,17,22].

Our research is a novelty due to the fact that the literature lacks the results of research on the biodrying of municipal waste on a full technical scale comparable to our scale, where in our reactor, there is 60 Mg of waste during the research. Moreover, there are no results of research with a similar scope of work to ours. According to the analysis presented in [19], the literature gives only four examples of research on biodrying on an industrial scale. Simultaneously, these are mostly the results from 2004–2009 and do not cover an important parameter such as the NHV of waste before, during, and after the process of biodrying. There are also no results of the NHV in [22], even though the research on biodrying was presented on a technical scale. That work focused on the material balance and the model estimation of water removal [22].

The main aim of this article was an analysis of biodrying, which was determined on the basis of the results obtained from a biodrying reactor working on a full industrial scale. The moisture, temperature, LOI, and NHV were tested during the process of the 14 day biodrying of waste.

2. Materials and Methods

2.1. Mechanical-Biological Treatment Plant

The results presented in this work were obtained from research conducted in the MBT installation in Końskie. This city is located in the center of Poland. The MSW collected from the 158,000-citizen community was the feedstock directed to the plant installations. The MBT plant works according to a two-stage technology where in the first stage, the mechanical operation by sieving (sieve 80 mm mesh) precedes the second biological stage (Figure 1). After the mechanical stage, the 0–80 mm fraction of the MSW undergoes a biodrying process in the reactor for a minimum of 14 days.

The reactor has external dimensions of 6.3 m × 6.7 m × 4.2 m and internal dimensions of 6.1 m × 6.11 m × 4.0 m. The reactor is equipped with a sectional overhead door for loading and unloading the reactor (Figure 2). The walls of the reactor have suitable insulation. The cuboid-shaped, galvanized steel reactor (150 m^3) is equipped with a module for active aeration connected with a bio-filter for removing odors. A set of 4 air supply ducts is installed on the bottom of the chamber, and an exhaust duct is mounted on the top of the chamber. The supply ventilator and the exhaust ventilator work continuously. Thus, during the processing in the drying reactor, the water vaporization is stimulated by a forced aeration system. The increasing temperature resulting from the activity of the microorganisms consequently leads to a gradually decreasing moisture in the waste. The reactor works in a periodic fashion and treats 50–60 Mg of waste at a time. The height of the layer of waste in the reactor is approximately 2.7 m. The reactor is equipped with temperature and moisture sensors (Figure 3).

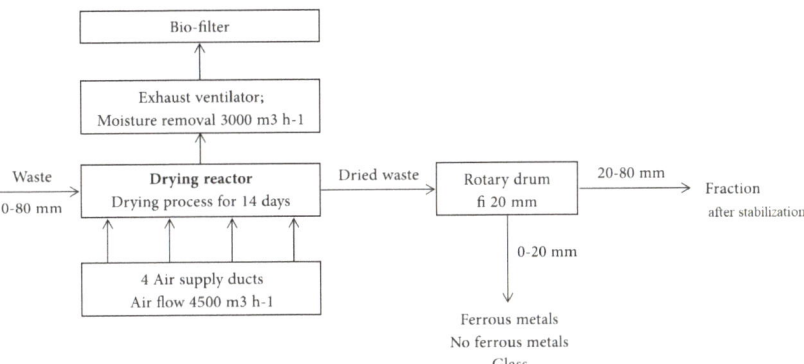

Figure 1. The scheme of the biological stage in the MBT installation.

Figure 2. The industrial reactor—the unloading of waste after 14 days of biodrying.

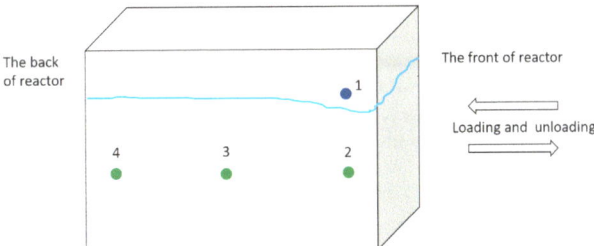

Figure 3. Schematic location of measurement points in the industrial reactor. 1—measurement point of the internal air; 2—measurement point in waste bed in the front; 3—measurement point in waste bed in the middle; 4—measurement point in waste in the back; blue line—the waste fill level of the reactor.

After the biodrying stage, the waste is transported to a rotary drum to remove the 0–20 mm fraction (ballast), as the 0–20 mm fraction is defined as a mineral, considered useless, and should be directed to the landfill site (Figure 1). According to the technology, the 20–80 mm fraction after biodrying and decreasing the moisture content could be considered a potential RDF utilized by thermal methods, provided that the organic carbon content guarantees a sufficiently high heat value.

During the 14 day biodrying, the average daily temperature of the ambient air was from 7.4 °C to 14.9 °C, while the average daily relative humidity of the ambient air was from 62.5% to 92.8% [23].

2.2. Composition of Raw MSW

The annual composition of raw MSW (fraction 0–80 mm) was investigated according to [24,25]. Four 100 kg samples were taken quarterly and, according to the standard in [25], were sieved on a 10 mm mesh-size sieve. The >10 mm fraction was examined for the mass share of plastic, glass, metals, mineral, and biodegradable paper, as well as textiles.

The samples investigated in the physicochemical tests were taken according to the procedure designed for solid recovered fuels [26].

2.3. Testing of the Waste Features in a Biodrying Process

During the 14 days of the biodrying of the waste in the reactor, the variability features in the processed waste were monitored. The temperature and moisture characteristics, both in the mass of the bed and in the air above the bed, were examined. The accuracy of the temperature sensor was +/−1.0 °C. Every waste sample taken from the drying reactor was tested (day by day).

After each day of the reactor operation, the collected waste samples were tested in order to evaluate parameters such as the moisture, LOI, and NHV.

The organic matter content (OM) can be expressed by the organic carbon content, which in the presented experiment was analyzed (in a certain simplification) by the LOI in the samples of the waste during the biodrying process. During the process of biodrying, organic matter undergoes partial decomposition by oxygen microorganisms. Thus, the progress of this process of decomposition can be analyzed with the differences of the LOI values in the subsequent days of the research. The samples for the determination of the LOI were taken from the front, middle, and back of the reactor. The average values of the samples taken after each day of the biodrying were used for the LOI calculations. The samples for the LOI determination were dried at 105 °C [27]. The tests of the LOI were conducted in a muffle furnace (type: M 104, Heraeus Instruments) by incineration at 550 °C for 6 h, using 5 g samples to calculate the average value, according to [28].

The NHV was determined in accordance with [29,30]. The heat of combustion determination was carried out in three replications. The tests for the heat of waste combustion assessment were obtained by mixing the ground laboratory samples obtained from the general tests from different parts of the reactor. Thus, one result of the heat of combustion was obtained for each day of the waste biodrying cycle in the reactor, without a division into the front, middle, or back of the reactor. In total, 42 results were obtained in this research. The tests were conducted with the use of the calorimeter (type: KL-12Mn, producer: Precyzja-BIT) with a measurement accuracy of the temperature increase of +/− 0.001 °C.

3. Results and Discussion

The biodegradable fraction contained in the waste directed to the reactor after summarizing the contribution of kitchen waste and paper was assessed at about 27.7%, but the contribution of the typical combustible waste such as multi-materials, plastic, foils, and fabric was assessed at about 17% (Table 1).

Table 1. The composition of the feedstock (0–80 mm fraction) directed to the reactor [31,32].

Components	Contribution *, % d.m.
Glass	18.0 ± 1.2
<10 mm Fraction	15.7 ± 0.7
Kitchen waste	16.6 ± 3.7
Mineral fraction	14.0 ± 4.8
Paper	11.1 ± 1.6
Multi-material materials	5.8 ± 0.2
Plastic	5.1 ± 0.2
Ceramic	4.9 ± 0.3
Foils	4.7 ± 0.02
Metals	2.4 ± 0.02
Fabrics	1.6 ± 0.03

* ±standard deviation.

The low level of organic waste should be noted as it is the result imposed by the requirements of the segregation at the source of the waste generation after 2012. The mineral fraction (glass, ceramic, mineral fraction, and metals) constituted about 39% of the share (Table 1).

The temperature was recorded in the reactor during the whole waste drying process, as shown in Figure 4. In the early days up to the 4th day, an intense temperature increase was observed, both in the mass of the waste and in the air above the bed. It should be noted that the temperature variation depends on where the temperature sensor is located. A temperature jump to around 70 °C occurred on the first day on the reactor's back wall. This tendency changed due to the reactor being opened every day for the sampling. The average temperature, that is representative of the waste bed is shown by the points for the middle batch of the reactor. These points represent just the highest mean temperatures in the whole cycle. In the following days up to the 9th day included, the temperature dropped to a level of about 30 °C and was then stabilized at this level.

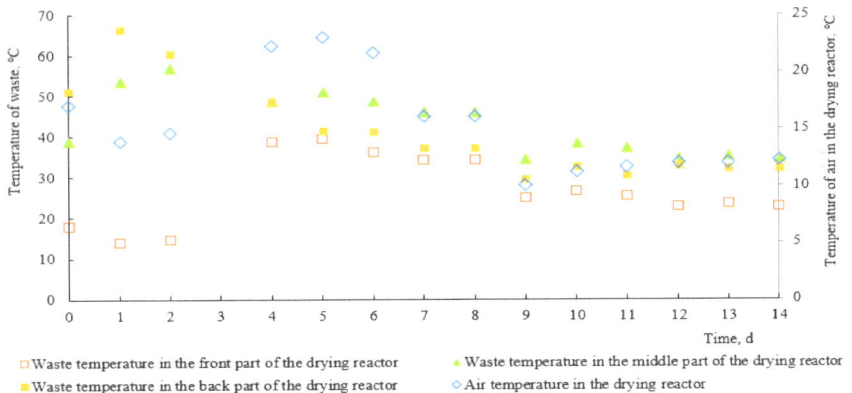

Figure 4. The temperature registered during the 14 days of the biodrying in the three parts of the reactor.

The average moisture content in the waste directed to the reactor was 56.5%. The variability in the moisture of the waste during the 14 days of processing in the reactor is presented in Figure 5. The recorded changes in the waste moisture in the drying process were regular and the moisture decreased from 56.5% to 30–35% (depending on the part of the biodrying reactor) in the final day of drying. Therefore, during the drying cycle, about 24% of the mass of the water was released into the atmosphere. Comparing this result of the drying effect to the result of 20% obtained by Hurka et al. [33], the drying effect in the analyzed reactor was comparable or even slightly better.

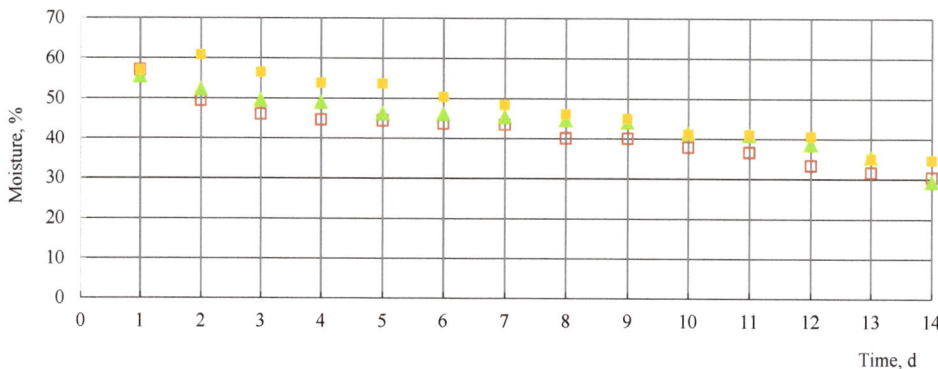

Figure 5. The moisture of the waste registered in the three parts of the drying reactor.

The water loss in the waste mass undoubtedly improved the calorific value of the waste. However, at the same time, a decrease in the share of organic carbon was observed, estimated here using the LOI (Figure 6), which is a consequence of the biomass decay. The value of the LOI of the waste directed into the reactor was about 70%. After 14 days of the biodrying process, the LOI loss was at about 33% to 18% (depending on the part of the mass location in the reactor). A similar reduction in the organic matter was noted by the authors in works [10,22]. During the 14 day biodrying cycle, the highest loss of the LOI to a level of about 37% occurred in the central part of the waste deposit. The lowest decrease in the LOI value to a level of about 52% was recorded in the front part of the reactor. The changes in the LOI parameter were reflected in the changes in the temperature parameter. The higher temperatures in the waste deposit favored the decomposition of the organic matter at a higher level.

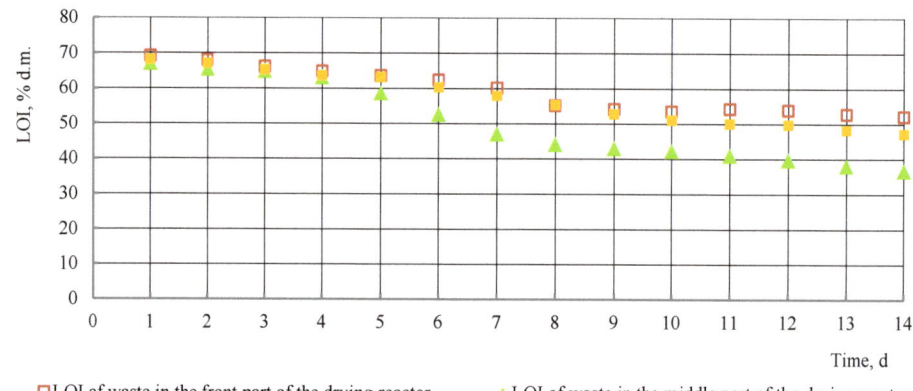

Figure 6. The loss on ignition of the waste registered during the 14 day cycle of the biodrying process in the different parts of the reactor.

As a result of the bio-drying, the NHV of the waste increased by more than 30% in comparison to the primary sample (Figure 7). A similar increase in the NHV of the waste (27%) on the pilot scale was presented by Negoi et al. [34]. However, the bio-drying process according to [34] was twice as long, whereas, according to [20], the biodrying of the waste at 55 °C during the 15 days caused an increase in the NHV of about 45% compared to the primary sample.

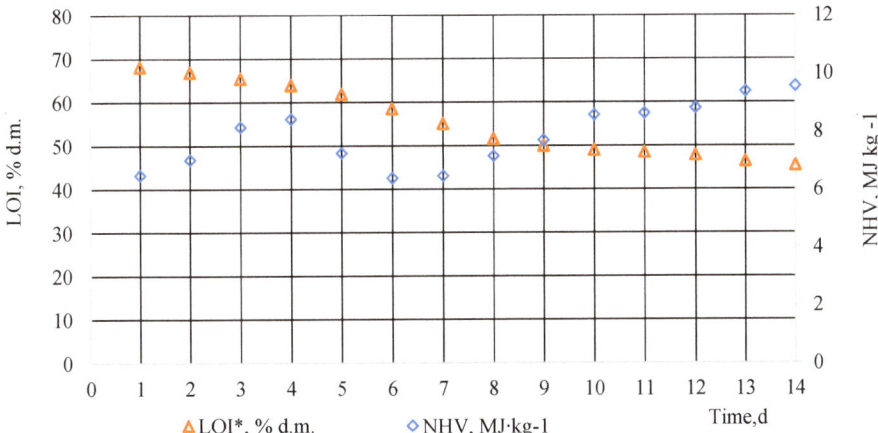

Figure 7. The net heating value and the loss on ignitions registered in the 14 day cycle of the biodrying process. * the LOI is the daily average value recorded from results investigated in the front, middle, and back part of the biodrying reactor.

The results give reason to conclude that, in terms of the NHV of the waste, it is important to properly control the temperature of the process inside the reactor, with minimal effect on the loss of carbon. The changes in the NHV measured in the waste during the 14 day cycle of the biodrying process allow us to confirm the obvious dependence that, along with the decrease in moisture, the NHV of the waste increased. The observed loss in the LOI was caused both by the decomposition of the organic matter as a result of the activity of microorganisms and the combustion of plastics during the standardized test.

4. Conclusions

The results of the research of biodrying in our case study under the industrial conditions on a full scale allow for the conclusions that in the 14 days of the process:

- The moisture of waste decreased to 50%.
- The carbon content expressed as the LOI decreased in waste to 66%.
- The NHV increased by more than 45%.
- The increase in the NHV was affected more by the decrease in moisture in waste than the change in the LOI. The loss of carbon was obtained as the effect of the process of biodegradation.

The results have application value as the plant operator, having at their disposal the controlling of the reactor's ventilation and the temperature inside the reactor, should select the speed of the moisture removal from MSW at such a level as to minimize the LOI and to keep the NHV at the highest level.

Author Contributions: Conceptualization, M.Ż.; methodology, M.Ż. and M.D.; formal analysis, J.L., M.Ż. and M.D.; investigation, M.D.; resources, M.D.; writing—original draft, M.Ż. and J.L.; writing—review and editing, J.L. and M.Ż.; visualization, J.L. All authors have read and agreed to the published version of the manuscript.

Funding: The project is supported by the program of the Minister of Science and Higher Education under the name: "Regional Initiative of Excellence" in 2019–2022 project number 025/RID/2018/19 financing amount PLN 12,000,000.

Institutional Review Board Statement: Not applicable.

Informed Consent Statement: Not applicable.

Data Availability Statement: Not applicable.

Acknowledgments: Investigations were led using: OXYMAX respirometer ER-10 (Columbus Instruments). Equipment was founded by the MOLAB Project in Kielce University of Technology: POIG 02.02.00-26-023/08-00; and project was supported by the Ministry of High Education.05.0.09.00/2.01.01.01.0004 MNSP.IKIO.14.001.

Conflicts of Interest: The authors declare no conflict of interest.

References

1. Council Directive 1999/31/EC of 26 April 1999 on the Landfill of Waste. Available online: https://eur-lex.europa.eu (accessed on 20 December 2021).
2. Directive 2008/98/EC of the European Parliament and of the Council of 19 November 2008 on Waste and Repealing Certain Directives. Available online: https://eur-lex.europa.eu/search.html?scope=EURLEX&text=Directive+2008%2F98%2FEC+of+the+European+Parliament+and+of+the+Council+of+19+November+2008+on+Waste+and+Repealing+Certain+Directives&lang=es&type=quick&qid=1643509255451 (accessed on 20 December 2021).
3. Waste Act (J.L. 2013, No. 0, Item. 21). Available online: https://isap.sejm.gov.pl/isap.nsf/download.xsp/WDU20130000021/O/D20130021.pdf (accessed on 19 December 2021).
4. Minister of the Economy. *Regulation of the Minister of the Economy of 16 of 16 July 2015 on the Acceptance of Waste for Landfilling (J.L.2015, No.0, Item. 1277)*; Minister of the Economy: Warsaw, Poland, 2015.
5. Montejo, C.; Tonini, D.; Marrquez, M.C. Mechanical-biological treatment: Performance and potentials, An LCA of 8 MBT plants including waste characterization. *J. Environ. Manag.* **2013**, *128*, 661–673. [CrossRef] [PubMed]
6. Dębicka, M.; Żygadło, M.; Latosińska, J. Investigations of bio-drying process of municipal solid waste. *Ecol. Chem. Eng. A* **2013**, *20*, 1461–1470.
7. Kasiński, S.; Dębowski, M.; Olkowska, M.; Rudnicki, M. Analysis of the Long-Term Mass balance and Efficiency of Waste Recovery in a Municipal Waste Biodrying Plant. *Energies* **2021**, *14*, 7711. [CrossRef]
8. Putranto, A.; Chen, X.D. A new model to predict diffusive self-heating during composting incorporating the reaction engineering approach (REA) framework. *Bioresour. Technol.* **2017**, *232*, 211–221. [CrossRef] [PubMed]
9. Ab Jalil, N.A.; Basri, H.; Ahmad Basri, N.E.; Abushammala, M.F.M. Biodrying of municipal solid waste under different ventilation periods. *Environ. Eng. Res.* **2016**, *21*, 145–151. [CrossRef]
10. Tambone, F.; Scagalia, B.; Scotti, S.; Adani, F. Effects of biodrying process on municipal solid waste properties. *Bioresour. Technol.* **2011**, *102*, 7443–7450. [CrossRef] [PubMed]
11. Shao, L.M.; Ma, Z.H.; Zhang, H.; Zhang, D.Q.; He, P.J. Biodrying and size sorting of municipal solid waste with high water content for improving energy recovery. *Waste Manag.* **2010**, *30*, 1165–1170. [CrossRef]
12. Contreras-Cisneros, R.M.; Orozco-Álvarez, C.; Piña-Guzmán, A.B.; Ballesteros-Vásquez, L.C.; Molina-Escobar, L.; Alcántara-García, S.S.; Robles-Martínez, F. The Relationship of moisture and temperature to the concentration of O_2 and CO_2 during biodrying in semi-static piles. *Processes* **2021**, *9*, 520. [CrossRef]
13. Gajewska, T.; Malinowski, M.; Szkoda, M. The Use of Biodrying to prevent self-heating of alternative fuel. *Materials* **2019**, *12*, 3039. [CrossRef]
14. Psaltis, P.; Komilis, D. Environmental and economic assessment of the use of biodrying before thermal of municipal solid waste. *Waste Manag.* **2019**, *83*, 95–103. [CrossRef]
15. Bilgin, M.; Tulun, S. Biodrying for municipal solid waste: Volume and weight reduction. *Environ. Technol.* **2015**, *36*, 1–23. [CrossRef] [PubMed]
16. Tom, A.; Haridas, A.; Pawels, R. Biodrying Process Efficiency: Significance of reactor matrix height. *Procedia Technol.* **2016**, *25*, 130–137. [CrossRef]
17. Vidriales-Escobar, G.; Rentiera-Tamayo, R.; Alatriste-Mondragon, F.; Gonzalez-Ortega, O. Mathematical modelling of a composting process in a small-scale tubular bioreactor. *Chem. Eng. Res. Des.* **2017**, *120*, 360–371. [CrossRef]
18. Tom, A.P.; Pawels, R.; Haridas, A. Biodrying process: A sustainable technology for treatment of municipal solid waste with high moisture content. *Waste Manag.* **2016**, *49*, 64–72. [CrossRef]
19. Tun, M.M.; Juchelková, D. Drying methods for municipal solid waste quality improvement in the developed and developing countries: A Review. *Environ. Eng. Res.* **2018**, *24*, 529–542. [CrossRef]
20. Evangelou, A.; Gerassimidou, S.; Mavrakis, N.; Komilis, D. Monitoring the performances of a real scale municipal solid waste composting and a biodrying facility using respiration activity indices. *Environ. Monit. Assess.* **2016**, *188*, 302. [CrossRef]

21. Ab Jalil, N.A.; Basri, H.; Ahmad Basri, N.E.; Abushammala, M.F.M. The potential of biodrying as pre-treatment for municipal solid waste in Malaysia. *J. Adv. Rev. Sci. Res.* **2015**, *7*, 1–13.
22. Ham, G.Y.; Matsuto, T.; Tojo, Y.; Matsuo, T. Material and moisture balance in a full-scale bio-drying MBT system for solid recovered fuel production. *J. Mater. Cycles Waste Manag.* **2020**, *22*, 167–175. [CrossRef]
23. Institute of Meteorology and Water Management—National Research Institute. Available online: https://dane.imgw.pl (accessed on 20 December 2021).
24. *PN–Z–15006*; Polish Standard. Investigation of Waste Morphology. Standards Association of Poland: Warsaw, Poland, 1993.
25. Jędrczak, A.; Szpadt, R. Determination of the Test Methodology of the Sieve, Morphological and Chemical Composition of Municipal Wastes. Zielona Góra. 2006. Available online: www.mos.gov.pl (accessed on 10 December 2021).
26. *PN–EN 15442:2011*; Solid Recovered Fuels—Methods for Sampling. Standards Association of Poland: Warsaw, Poland, 2011.
27. *PN–93/Z15008*; Polish Standard. Fuel Property Testing. Determination of the Total Moisture. Standards Association of Poland: Warsaw, Poland, 1993.
28. *PN–EN 15169:2007*; Polish Standard. Characterization of Waste—Determination of Loss on Ignition in Waste, Sludge and Sediments. Standards Association of Poland: Warsaw, Poland, 2007.
29. *PN–93/Z–15008*; Polish Standard. Solid Municipal Wastes. Tests on the Fuel Properties. The Determination of Combustion Heat and the Calculation of the Calorific Value. Standards Association of Poland: Warsaw, Poland, 2004.
30. *ISO 1928:2009*; Solid Mineral Fuels. Determination of Gross Calorific Value by the Bomb Calorimetric Method and Calculation of Net Calorific Value. ISO: Geneva, Switzerland, 2009.
31. Dębicka, M.; Żygadło, M.; Latosińska, J. The effectiveness of biodrying waste treatment in full scale reactor. *Open Chem.* **2017**, *15*, 67–74. [CrossRef]
32. Żygadło, M.; Dębicka, M.; Latosińska, J. The bioconversion of municipal solid waste in the biodrying reactor. *J. Environ. Earth Sci.* **2019**, *1*, 33–37. [CrossRef]
33. Hurka, M.; Malinowski, M. Assessment of the use of EWA bioreactor in the process of biodrying of undersize fraction manufactured from mixed municipal solid waste. *Infrastruct. Ecol. Rural. Areas* **2014**, *4*, 1127–1136.
34. Negoi, R.M.; Ragazzi, M.; Apostol, T.; Rada, E.C.; Marculescu, C. Biodrying of Romanian municipal solid waste: An analysis of its viability. *UPB Sci. Bull. Ser. C* **2009**, *71*, 193–204.

Article

Immobilization of Zn and Cu in Conditions of Reduced C/N Ratio during Sewage Sludge Composting Process

Aleksandra Leśniańska, Beata Janowska * and Robert Sidełko

Faculty of Civil Engineering, Environmental and Geodetic Sciences, Koszalin University of Technology, 75-453 Koszalin, Poland; lesnianska@gmail.com (A.L.); robert.sidelko@tu.koszalin.pl (R.S.)
* Correspondence: beata.janowska@tu.koszalin.pl

Abstract: In this paper we present results of research on the transformation of chemical forms of two elements (Cu, Zn) that occurred at the highest concentration in sewage sludge being processed in a composting process. The factor that had impact on the direction of the observed transformation was the amount of straw added to the mix with sewage sludge at the batch preparation stage including elimination of an additional source of organic carbon (straw). The analysis of contents of Cu and Zn chemical forms was performed applying Tessiere's methodology. It was ascertained that reduction of supplementation has positive impact on the allocation of tested elements in organic (IV) and residual (V) fractions with a simultaneous decrease of heavy metals mobile forms share in bioavailable fractions, mostly ion exchangeable (I) and carbonate (II). Using an artificial neural network (ANN), a tool was developed to classify composts based on Austrian standards taking into account only I ÷ IV fractions treated as a labile, potentially bioavailable, part of heavy metals bound in various chemical forms in compost. The independent variables that were predictors in the ANN model were the composting time, C/N, and total content of the given element (total Cu, Zn). The sensitivity coefficients for three applied predictors varied around 1, which proves their significant impact on the final result. Correctness of the predictions of the generated network featuring an MLP 3-5-3 structure for the test set was 100%.

Keywords: composting; heavy metals; immobilization; speciation

Citation: Leśniańska, A.; Janowska, B.; Sidełko, R. Immobilization of Zn and Cu in Conditions of Reduced C/N Ratio during Sewage Sludge Composting Process. *Energies* **2022**, *15*, 4507. https://doi.org/10.3390/en15124507

Academic Editors: Robert Oleniacz and Katarzyna Grzesik

Received: 11 May 2022
Accepted: 17 June 2022
Published: 20 June 2022

Publisher's Note: MDPI stays neutral with regard to jurisdictional claims in published maps and institutional affiliations.

Copyright: © 2022 by the authors. Licensee MDPI, Basel, Switzerland. This article is an open access article distributed under the terms and conditions of the Creative Commons Attribution (CC BY) license (https://creativecommons.org/licenses/by/4.0/).

1. Introduction

Composting is a method of biological waste treatment, which guarantees, in the case of municipal waste, production of a biologically stable product, which is important in the context of its further management, e.g., by disposal [1]. Annually, 12.8 million tons of household and commercial waste are produced in Poland, whereof over 1.2 million tons, approximately 9.5%, are being processed through composting [2]. The composting process is also used to process sewage sludge originating from municipal wastewater treatment plants; its volume systematically increases. Comparing the data for 2000 and 2018, the volume of sewage sludge amassed in Poland increased by 62%, reaching 583 thousand tons of dry mass [3]. Being a method of biological transformation of biodegradable waste, composting features high dynamics of organic matter transformation due to mineralization and humification, which guarantees production of a product of high fertilizing values [4,5]. Compost manufactured based on municipal sewage sludge is qualified as a soil improver or an alternative substrate (growing media) used in industrial plant production [6,7]. High contents of macroelements cause fertilization with compost to increase the content of organic substances, thus improving soil properties. Unfortunately, due to increased concentrations of heavy metals in sewage sludge, there is a high risk of exceeding their admissible values in compost, thus limiting the possibility of its use [8].

A consequence of systematic use of compost with increased heavy metal concentrations to fertilize soil is the accumulation of microelements in the food chain, thus posing a considerable hazard for human health [9,10]. It should be underlined that in regulations applicable in various countries, maximum admissible values are indicated, defined by the total contents of heavy metals made up of the sum of all chemical forms of a given element (Table 1).

Table 1. Heavy metals limits in selected EU countries. Source: self-elaboration based on [1] Regulation (EU) 2019/1009 [11]; [2] Biala and Wilkinson [12].

Country	Concentration, mg·kg^{-1}d.m.							
	As	Cr	Cu	Hg	Ni	Pb	Zn	Cd
Poland [2]	-	100	-	2	60	140	-	5
The Netherlands [2]	15	50	90	0.3	20	100	290	1
Sweden [2]	-	100	100	1	50	100	300	-
France [2]	25	120	300	2	60	180	600	3
Greece [2]	15	510	500	5	200	500	2000	10
Germany [2]	1	100	100	1	50	150	400	1.5
Denmark	25	-	1000	0.8	30	120	4000	25
EU [1]	40	2	-	1	50	50	-	2

Using a sequential chemical extraction method such as the BCR (Community Bureau of Reference)-four-step extraction, or the five-step extraction method developed by Tessiere, comprehensive information on the distribution of heavy metals fractions can be obtained; this is useful for prediction of metals' mobility, bioavailability and leaching rates [13,14]. Speciation testing consisting of sequential extraction of heavy metals from compost samples by dissolution of various elements' chemical forms [15] clearly indicated that part of micro-pollutants can be considered as permanently bound within the medium mineral matrix [16–19].

According to available knowledge the correct course of composting in industrial conditions requires specific technological parameters, including the proportion between total organic carbon (TOC) and total nitrogen (TN). The optimum C/N ratio value was not fixed unequivocally and depending on the raw material type as well as the remaining parameters, i.e., humidity, intensity of aeration, and temperature, generally falls within the 25–35 interval [20]. When the composting process is performed, due to a high concentration of nitrogen in sewage sludge being 2–7% d.m. [21–23], supplementation consisting in addition of material increasing the organic carbon share in the composting mass is used [24–26].

Usage of various types of supplements that act as additional components modifying the composting mass composition as a factor regulating the required C/N parameter value was, and still is, a topic of many research works [27–30]. Considering the risk of origination of gaseous ammonia (NH_3) that has a toxic impact on microorganisms, in the case of increased organic nitrogen concentration [27,31], research on the composting process in which C/N values are lower than those recommended is relatively rare. A good example can be the research work performed by Kulikowska and Sindrewicz [32], who used a reactor of 1.3 m^3 volume and a relevant proportion between dehydrated sewage sludge and barley straw modifying the C/N parameter to attain a 15 value. During the said research work no negative impact of a decreased supplement in the composted mix on the process course was found. Similar conclusions were drawn from tests that were performed at an industrial scale, where the initial C/N ratio value in the batch amounted to 12 [33].

Although composting research pertaining to heavy metals speciation indicates that with the passage of time their concentrations in bioavailable fractions decreases and increases in inert fractions [34], the impact of limited supplementation decreasing C/N values on the direction of heavy metals' chemical forms' transformation has not been made clear. Meanwhile, the most frequently used supplement, i.e., straw, is a scarce material, the increasing market prices of which raise compost production costs; therefore, research works were commenced to verify the following thesis: (i) in the case of sewage sludge composting, any decrease in supplementation, including its elimination, has no negative impact on heavy metals' chemical forms' transformation towards inert forms, and (ii) increased nitrogen concentration decreasing the C/N value does not limit the composting process intensity.

During the commenced research, a change of the selected heavy metals' contents in those fractions that were separated via sequential extraction in compost samples produced from sewage sludge in decreased supplementation conditions was analyzed. The objective of the research work was evaluation of (i) the direction of transformation of zinc and copper chemical forms during composting depending on the batch quality made up of a mix of sewage sludge and straw fixing the C/N value, and (ii) a possibility to use an artificial neural network (NN) as a tool for compost classification taking into account those heavy metals' chemical forms that might pose a real hazard. The results of heavy metals' content analysis for sewage sludge used in the research work show that except for Zn and Cu, concentrations of the remaining elements (Cr, Cd, Ni, Pb, and Hg) did not exceed their admissible values [35], thus justifying limiting the speciation testing range to those two elements.

2. Material and Methods

2.1. Composting Experiment (Device)

Tests were performed on an industrial scale at the Goleniów (Poland) wastewater treatment plant within a project financed under the EU South Baltic [35] project. During the tests the course of mechanically dehydrated sewage sludge with structural materials added in various proportions was monitored. At stage I, a mix of sewage sludge with added barley straw, wood chips, and mature compost (inoculum) in mass proportions of 4:1:0.5:0.5 w/w (E1) was used. At stage II, the mix composition using the same components was 8:1:1:1 w/w (E2). At stage III, straw was abandoned and sewage sludge was mixed only with wood chips in 1:1 w/w proportion (E3) (Table 2).

Table 2. The composition of the mixture to be composted.

Mixture Components	E1	E2	E3
	Mass Proportion		
Sewage sludge	4	8	1
Barley straw	1	1	-
Wood chips	0.5	1	1
Mature compost (inoculum)	0.5	1	-

The composting process carried out in stages I and II was performed under roofed windrows featuring a trapezoid cross-section of approximately 70 m in length. These windrows were mechanically turned over twice per week during the first three weeks of composting, whereas in subsequent weeks, once per week on average. The composting process performed in stage III was done in similar windrows as in stages I and II but under GORECover® semi-permeable membranes in intense aeration conditions by pumping air through ducts located in the reactor concrete floor. The mass of each windrow was approximately 200 tons. Each of those three stages was repeated twice. Physical and chemical characteristics of the raw materials and composting initial mixtures are shown in Table 3.

Table 3. Physical and chemical characteristics of the composting initial mixtures used.

Material	Stage	Dry Mass	Organic Matter	Total Organic Carbon	Total Nitrogen	C/N	Cu	Zn
		[%]					[mg·kg^{-1} d.m.]	
Sewage sludge	E1	21.4	81.4	33.92	7.48	5.0	225.25	551.75
	E2	15.6	78.2	40.4	7.54	5.0	252.50	497.25
	E3	14.0	81.0	33.20	6.81	4.9	216.75	531.50
Barley straw	E1	62.4	94.2	45.7	0.7	66	3.50	12.75
	E2	62	94.0	45.5	0.71	64	3.60	11.82
Wood chips	E1	34.7	87.0	44.4	1.14	39	-	-
	E2	34.0	87.0	44.4	1.18	38	-	-
	E3	34.2	87.1	44.4	1.12	39	-	-
Compost (at 1 day)	E1	21.7	83.65	41.35	3.05	14.56	144.25	358.75
	E2	20.5	80.30	39.08	4.07	9.61	184.25	418.50
	E3	37.30	74.50	36.90	4.01	9.20	58.50	175.75

2.2. Physical and Chemical Parameters Analyses

During the field tests, temperature change in all composted windrows was monitored. From each windrow, five compost samples of approximately 1 kg in weight were taken, and after their mixing a sample for lab tests, according to the Polish standard PN-R-04006: 2000, was taken [36].

Particular parameters of compost samples were determined using mainly standard research methods in accordance with Polish standards (PN). The lab tests comprised determination of dry matter content (d.m.) after sample drying at 105 °C (PN-R-04006), the organic matter content was determined by the loss on ignition of the dry mass at 550 °C (PN-Z-15011-3), and total organic carbon concentration (TOC)—PN-Z-15011-1,3 and total nitrogen (TN)—PN-R-04006 were determined using a Vario MAX CN analyzer [33]. Samples for the analysis were prepared in accordance with the methodology described in Polish standard PN-Z-15011-3: 2001 [37].

Cu and Zn fractionation was performed using Tessier's modified sequential extraction (Table 4). Tested metals' contents in particular fractions were determined using flame atomic absorption spectrometry (FAAS)-iCE 3500Z THERMO SCIENTIFIC. Total Cu and Zn contents were determined in the same way as fraction V.

Table 4. Analytical procedure (adapted with permission from [34]).

Fraction	Extractant	Extraction Conditions	
		Temperature	Time
FR I Exchangeable	10 cm^3 1 M CH$_3$COONH$_4$ pH = 7	20 °C	1 h
FR II Carbonate	20 cm^3 1 M CH$_3$COONa, pH = 5	20 °C	5 h
FR III Bound with Mn and Fe oxides	20 cm^3 0.04 M NH$_2$OH·HCl w 25% (v/v) CH$_3$COOH	95 °C	5 h
FR IV Organics and sulphides	(a) 5 cm^3 0.02 M HNO$_3$ + 5 cm^3 30% H$_2$O$_2$, pH = 2 (b) 5 cm^3 30% H$_2$O$_2$, pH = 2 (c) 10 cm^3 3.2M CH$_3$COONH$_4$ w 20% (v/v) HNO$_3$	(a) 85 °C (b) 85 °C (c) 20 °C	2 h 3 h 0.5 h
FR V Residue	5 cm^3 65% HNO$_3$ + 1 cm^3 30% H$_2$O$_2$ + 1 cm^3 75% HClO$_4$	Microwave mineralization	

2.3. Statistical Analysis

Mobility of elements is defined as an ability to transfer from a sample solid phase with which a given element form is weakly bound, and which can be liberated in natural conditions (e.g., ionic form of carbonates). To define mobility and bioavailability factors of bioavailability (MF) [38] are used. MFs of Cu and Zn were defined as the ratio of the metal content in FI and FII to the total content.

The inactivation rate (*IR*) of heavy metals was calculated as follows [10,13]:

$$IR(\%) = \frac{(R_b - R_a)}{R_b} \times 100, \quad (1)$$

where *IR* is the passivation of the heavy metal (%), R_b is the distribution ratio for the exchangeable fraction (FI + FII) of the heavy metal before composting (%), and R_a is the distribution ratio for the exchangeable fraction (FI + FII) of the heavy metal after composting (%).

The distribution ratio (*R*) for the exchangeable fraction was calculated using:

$$IR(\%) = \frac{C_e}{C_f} \times 100, \quad (2)$$

where C_e is the exchangeable content of the heavy metal (mg·kg^{-1}), i.e., the sum of metals' content present in the FI and FII fractions, and C_f is the total content of the heavy metal (mg·kg^{-1}).

For statistical analysis of test results, Microsoft EXCEL software, 2007 version, and STATISTICA of StatSoft version 13.1 (Campus-Wide License) were used. The scope of applied statistical tools comprised the nonlinear regression issues and selected statistics available in the basic statistics and tables module of STATISTICA. The numerical analysis was performed using the Neural Networks 8 PL module of STATISTICA applet-StatSoft.

3. Results and Discussion

3.1. Statistical Analysis

3.1.1. Change of Temperature, Organic Matter Degradation, Total Organic Carbon, and Total Nitrogen

Temperature is strongly correlated with a reaction's biological velocity; therefore, it is frequently used to reflect the activity of microorganisms and to define composting stability [14]. In compost windrows featuring different initial C/N ratios, a sharp increase of temperature occurred on the third and fourth days of composting followed by a thermophilic phase, which lasted until approximately the 35th day (Figure 1a). Windrows featuring lower C/N ratios (9.20 and 9.61) manifested higher maximum temperature values. At the beginning of the composting process with an initial C/N ratio of 14.56, the temperature increase rate was slower than that observed in the composting process for C/N ratios of 9.20 and 9.61. The highest temperature values were noted in E3 (C/N 9.20), which could be a result of application of the windrow prism cover. The lowest average temperature values were recorded in the windrows at stage 1. In this stage the thermophilic phase was the shortest. Temperature values' distribution had a different course than in the research work of Wu et al. [14] pertaining to a pig manure composting process. The temperature increase rate was lower in those composts that had the lowest C/N (12.5 and 15) ratio values.

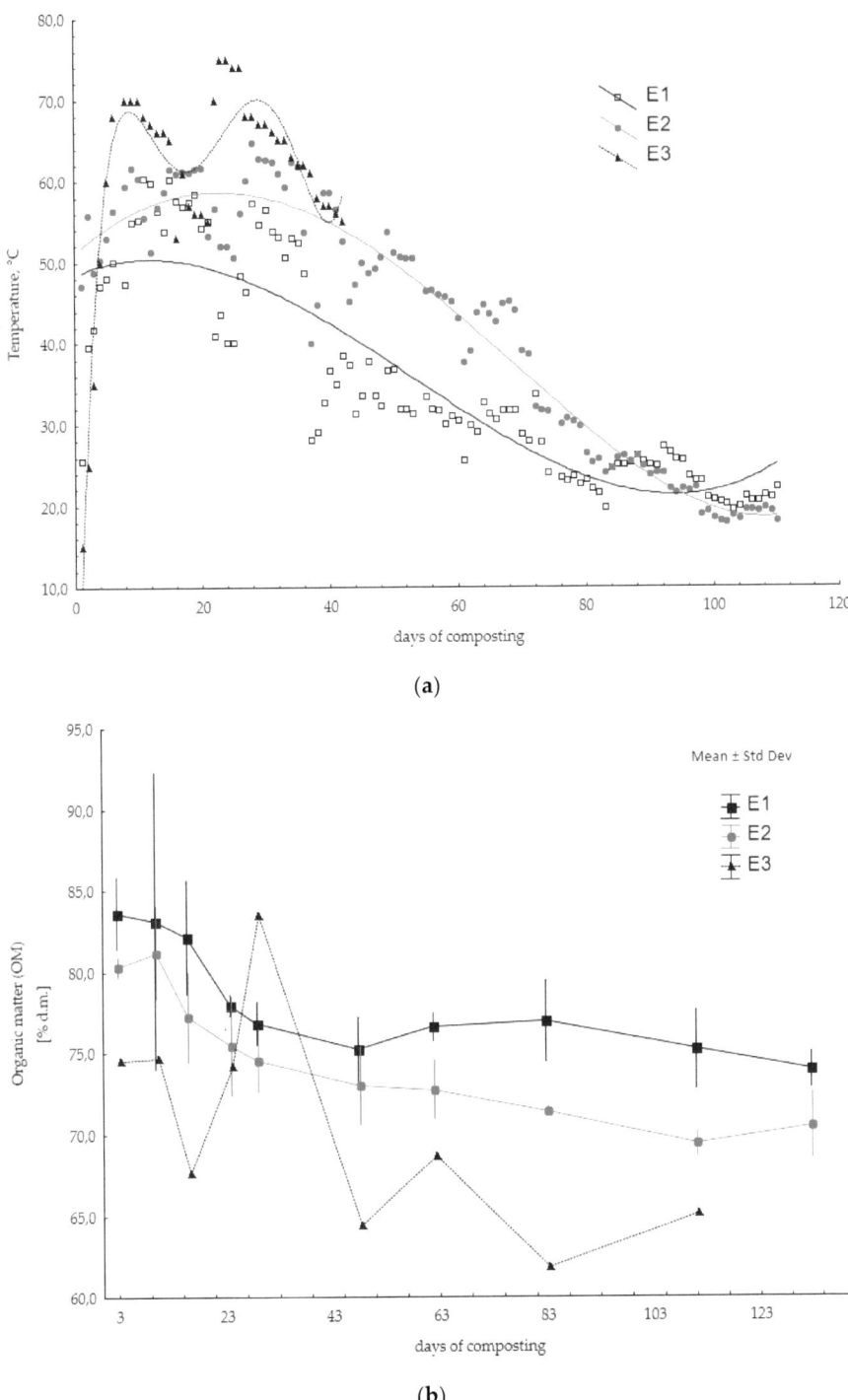

Figure 1. *Cont.*

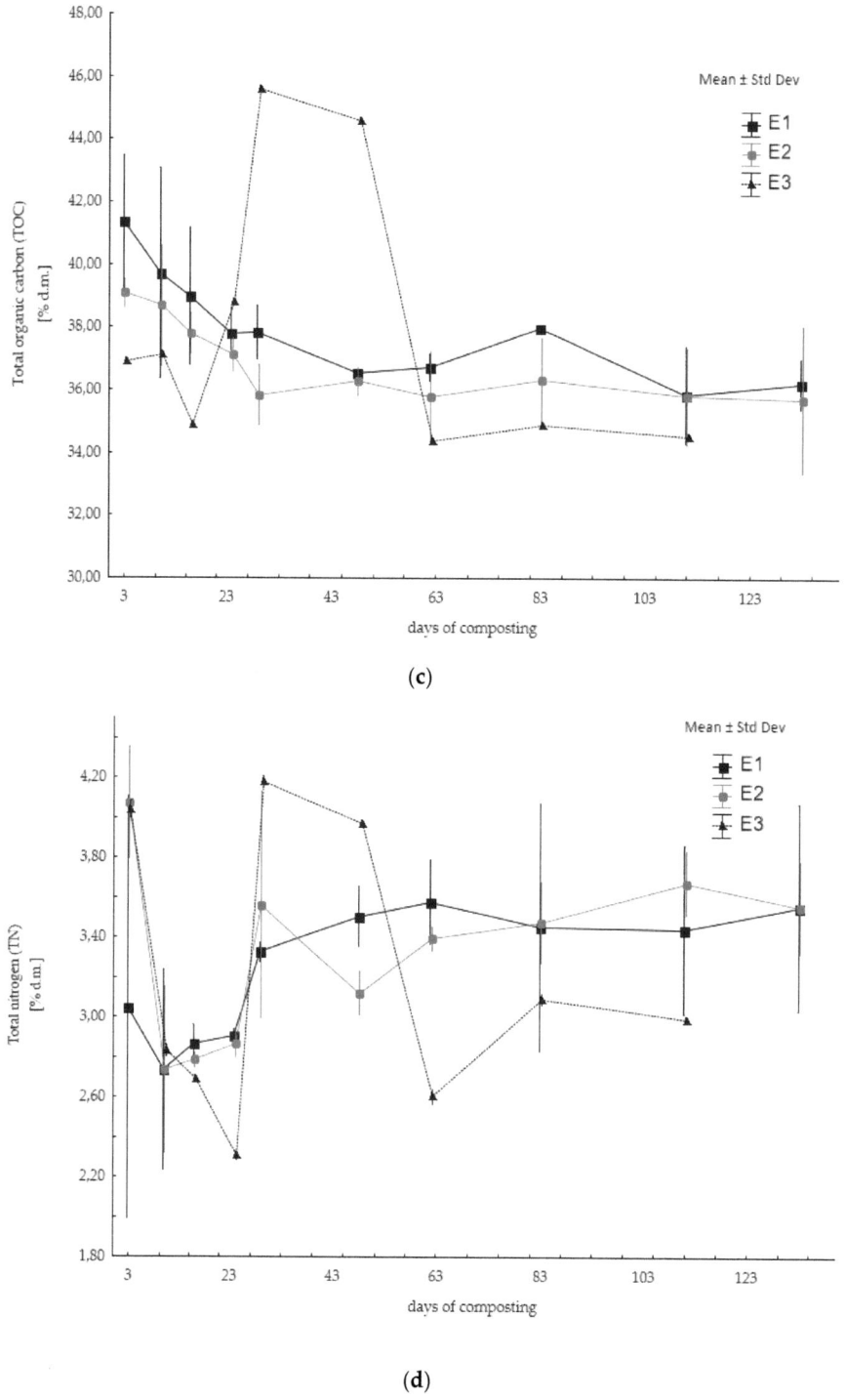

Figure 1. Changes in temperature (**a**), OM (**b**), TOC (**c**), and TN (**d**) during composting.

Organic matter content (OM) decreased in all three stages during the composting process (Figure 1b). The highest organic matter content at the initial composting phase featured the samples from the first composting stage (E1)—83.65%, whereas the lowest OM content value was noted for stage 3 compost—74.50%. In the last day of the composting process, OM content values were 73.95% (E1), 70.50% (E2), and 65.10% (E3). In E2 and E3, the highest OM loss was noted after the 16th day of the composting process, whereas in E1, it was after the 24th day. OM losses can be associated with the chemical composition of composted materials, particularly with fiber concentration. In compost windrows featuring different initial C/N ratios, loss of OM during the composting process was comparable [13,14].

Change of total organic carbon (TOC) contents during the composting process manifested a similar pattern. The highest TOC value at the process beginning was noted in stage 1 windrows, at 41.35%, and the lowest value at stage 3, at 36.90% (Figure 1c). At the initial phase of the composting process the lowest total nitrogen (TN) value was noted for sewage sludge that had the highest initial values of C/N ratio (14.56), at 3.05% (E1). In those sewage sludge mixes with a reduced volume of straw (E2) and/or without (E3), initial TN content was 4.07% and 4.01%, respectively. TN content values noted during the composting thermophilic phase (up to the 24th day) decreased, which could be caused by liberation of NH_4^+ ions and emission of gaseous ammonia (NH_3), which could increase odor onerousness (Figure 1d). At the final composting phase in those windrows that had the highest C/N ratio values, a slight increase of TN contents were noted. A slow increase of nitrogen concentration in compost at its maturity phase should be considered as beneficial due to the fertilizing values of this element [1,14,33].

3.1.2. Zinc and Copper Speciation

The average total copper content in sewage sludge used for composting purposes in E1, E2, and E3 amounted to 225.25 mg kg^{-1} d.m., 252.25 mg kg^{-1} d.m., and 216.75 mg kg^{-1} d.m. respectively. Total Cu content in tested compost samples taken from E1, E2, and E3 windrows increased respectively from 144.25 mg kg^{-1} d.m., 184.25 mg kg^{-1} d.m., and 58.50 mg kg^{-1} d.m. to 209.38 mg kg^{-1} d.m., 239.83 mg kg^{-1} d.m., and 122.50 mg kg^{-1} d.m. Total Cu content in produced composts did not exceed its admissible values (300 mg kg^{-1} d.m.) defined for organic fertilizer and soil improvers, which pertain to putting fertilizers on the market [11]. The highest Cu content increase was noted for E3, where Cu content increased 2.1-fold compared with the initial content, and the lowest increase of total Cu content (1.3-fold) was noted for E2. Similar results were noted in many research works and such an increase has been attributed to the loss of mass caused by organic matter decomposition, CO_2, and water liberation, as well as mineralization processes [13,14,19]. The changes of copper compounds contents in particular fractions taken from windrows at the E1, E2, and E3 stages are presented in Figure 2a–e.

In all windrows featuring different initial C/N ratio values, the highest concentrations were noted for Cu compounds bound with the organic matter (FV). The results were in line with the well-known Cu affinity to organic substances, which contributed to the generation of humic substances and confirmed high Cu affinity to the –OH or –COOH humic substances functional groups [39]. The highest Cu content in particular fractions was noted for samples taken from the windrow of E2 whereas the lowest Cu contents were noted in samples taken from E3, which was influenced by the initial Cu content.

Copper in tested samples was mainly bound with organic matter (FIV). The Cu compounds' percentage share in this fraction, at all stages, was within 59.56–77.60% of the total value interval. During composting performed in windrows of C/N 14.56 and 9.61 (E1, E2), an increase of the Cu percentage in FIV was noted (Table 5). Ion exchangeable Cu compounds liberated through OM degradation are bound by phenol and quinone groups of originating humic acids confirming high Cu affinity to the –OH and –COOH functional groups [19,39]. Origination of humic acids during the composting process has a significant impact on the limitation of Cu bioavailability [18,39].

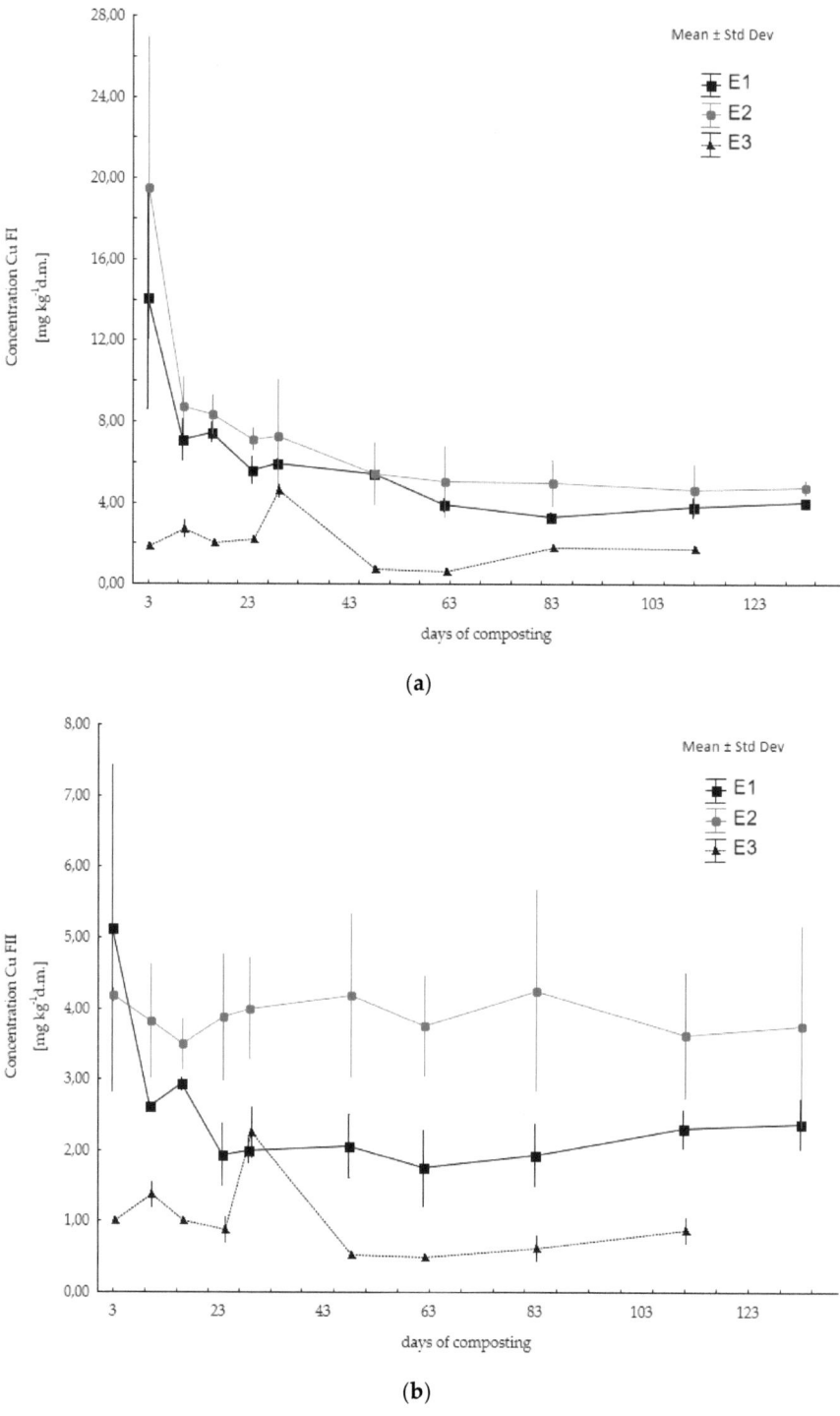

Figure 2. *Cont.*

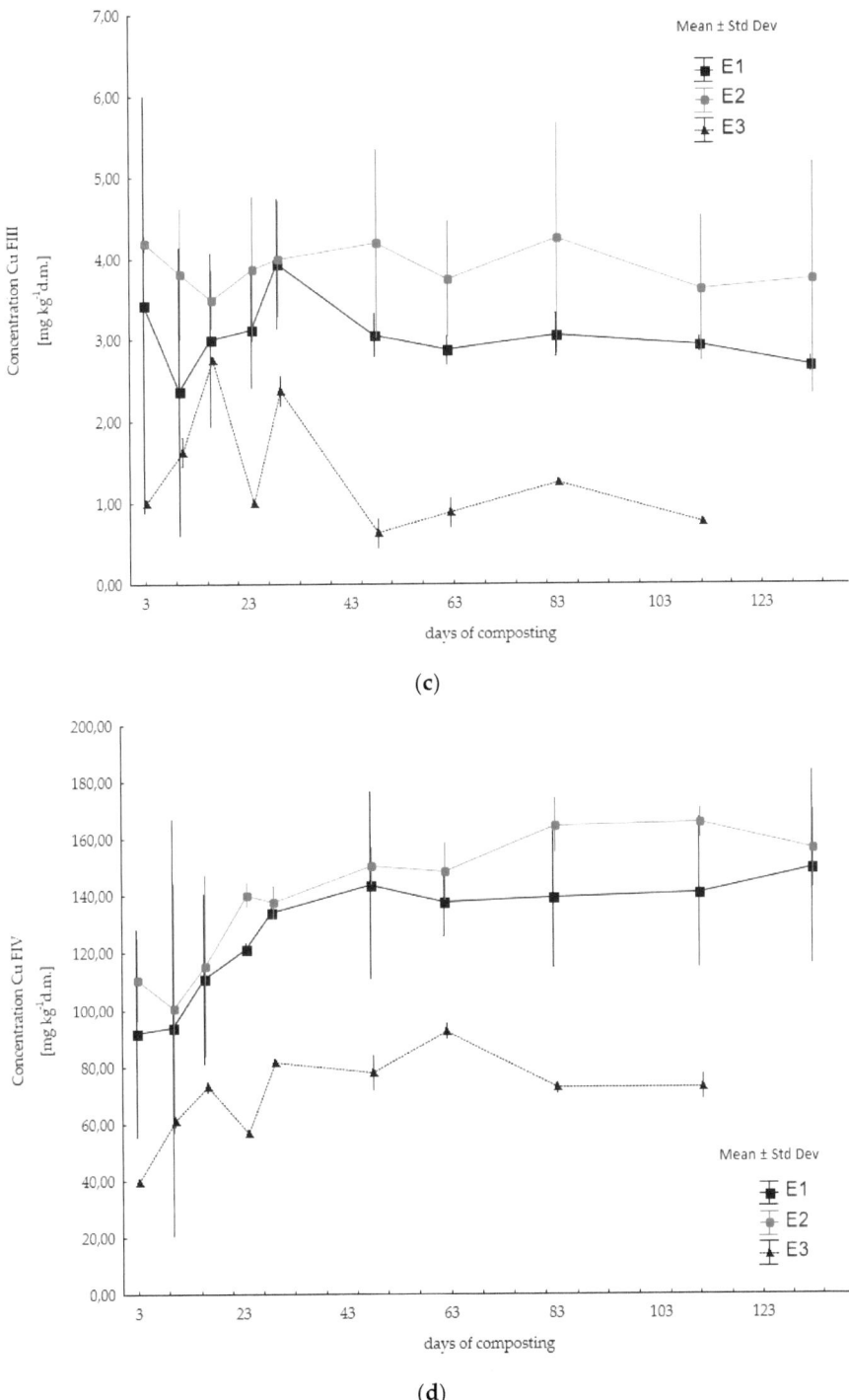

Figure 2. *Cont.*

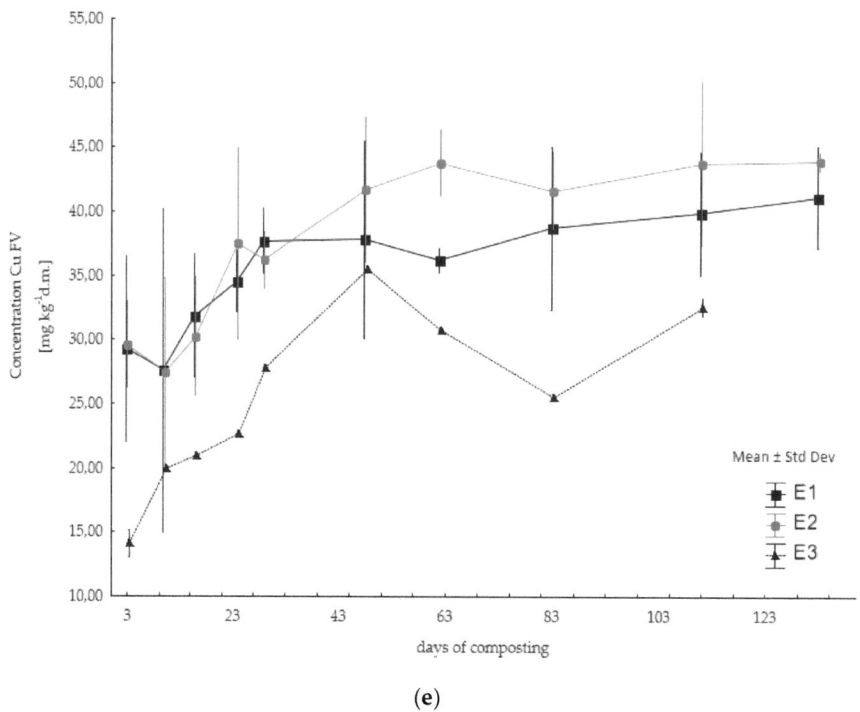

(e)

Figure 2. Concentration of Cu in particular fractions FI (**a**), FII (**b**), FIII (**c**), FIV (**d**), and FV (**e**) in stages E1, E2, E3.

Table 5. Increments of Cu and Zn fractions during composting (end vs. start. %) for composts with different initial C/N and inactivation rate (IR) values.

Element	Stage	FI	FII	FIII	FIV	FV	MF	IR
Cu	E1	−7.86	−2.37	−0.60	4.80	−5.19	−10.03	77.74
	E2	−8.79	−2.02	−0.65	6.57	2.63	−11.51	80.31
	E3	−1.76	−1.43	−1.18	−8.01	10.25	−2.38	61.31
Zn	E1	−1.19	−13.12	0.62	15.93	−0.61	−16.45	41.66
	E2	−2.01	−14.26	3.05	9.39	−5.58	−16.74	50.31
	E3	−6.49	−8.78	−4.00	16.15	0.39	−17.23	51.91

Cu percentage in the V fraction varied from 16.38% to 27.51%. The lowest Cu percentage was noted in fractions III and II, making approximately 2% of the total content. Ion exchangeable water soluble Cu compounds varied from 0.41% do 3.46% of the total content.

The lowest Cu percentage in fractions I, II, III, and IV was noted for compost samples taken at the E3 stage. This compost featured the highest Cu compounds percentage in fraction V. Compost samples taken at the E2 stage featured the highest Cu percentage in fraction III, whereas sewage sludge composted with the highest straw content (E1) had the highest Cu percentage in fractions I, II, and IV.

Comparing the obtained test results with those presented by Xu et al. [13], the sum of fractions I and II at the initial composting phase was similar (13.16%), whereas Cu distribution in the remaining fractions differed considerably.

Average total zinc content in sewage sludge used for composting at the E1, E2, and E3 stages amounted to 551.75 mg kg^{-1}d.m., 497.25 mg kg^{-1}d.m., and 531.50 mg kg^{-1}d.m., respectively. Similar results were obtained in other studies in which the increase in OM was attributed to weight loss due to microbial degradation [14,19].

In tested compost samples of the 1, 2 and 3 stages, total Zn content increased from 358.75 mg kg^{-1}d.m. to 465.70 mg kg^{-1}d.m., 418.50 mg kg^{-1}d.m. to 520.08 mg kg^{-1}d.m., and 175.75 mg kg^{-1}d.m. to 294.50 mg kg^{-1}d.m., respectively. Total Zn content in produced composts did not exceed the admissible contents (800 mg kg^{-1}d.m.) fixed for organic fertilizer and soil improvers [11]. The highest Zn content increase was noted in E3, where Zn content increased 1.7-fold compared with the initial content, whereas in E1 and E2 Zn content in compost samples taken at the terminal process phase increased 1.3- and 1.2-fold, respectively. Change of zinc compounds contents in particular fractions taken from windrows at the E1, E2, and E3 stages are presented in Figure 3a–e.

The highest Zn contents in fractions II, III, IV, and V were noted for samples taken from the windrow at stage 2. The highest concentrations of soluble and ion exchangeable Zn compounds at the thermophilic composting phase were noted for compost samples taken at stage 2 (E2). Composted sewage sludge without straw added (E3) had the lowest Zn contents in all fractions. Zn redistribution in particular fractions differed from Cu redistribution. Zn in the tested samples was bound mainly with Fe/Mn oxides (FIII)—from 29.40% to 47.86%, organic matter (FIV)—from 14.57% to 40.78%, and carbonates (FII)—from 12.93% to 31.64%. The lowest Zn percentage was noted for fractions I and V. Ion exchangeable Zn compounds' contents made up from 0.49% to 7.81% of the total Zn content. Zn percentage in the residual fraction varied from 2.77% to 15.43%. Similar results concerning the distribution of Zn in the next five fractions in composted sewage sludge were obtained by Wang et al. [9]. The percentage of Zn compounds bound to Fe/Mn oxides was higher by 40% to 56.9%, while Zn content in the organic fraction decreased by approx. 7.7% [19].

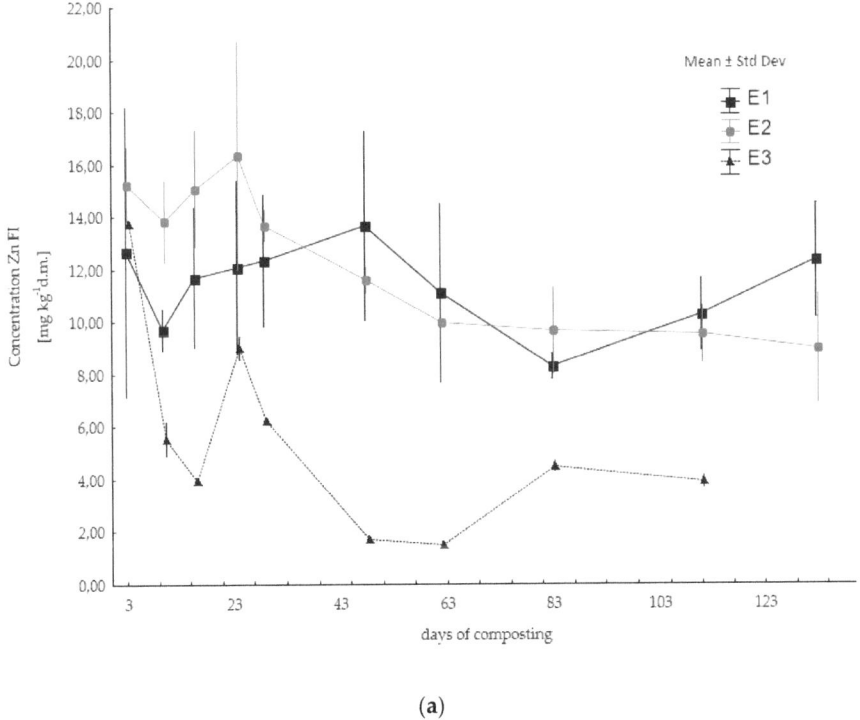

(a)

Figure 3. Cont.

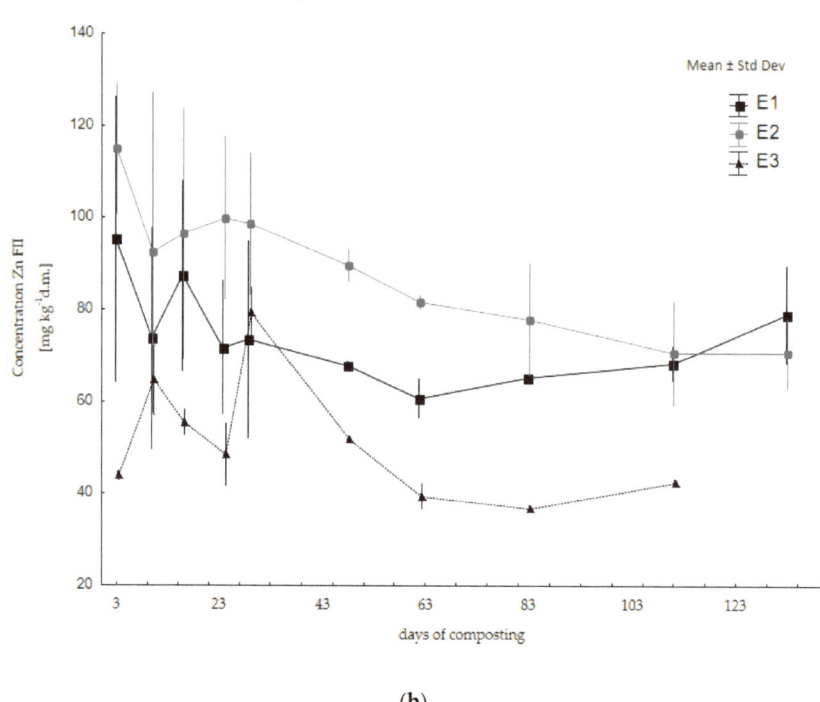

(b)

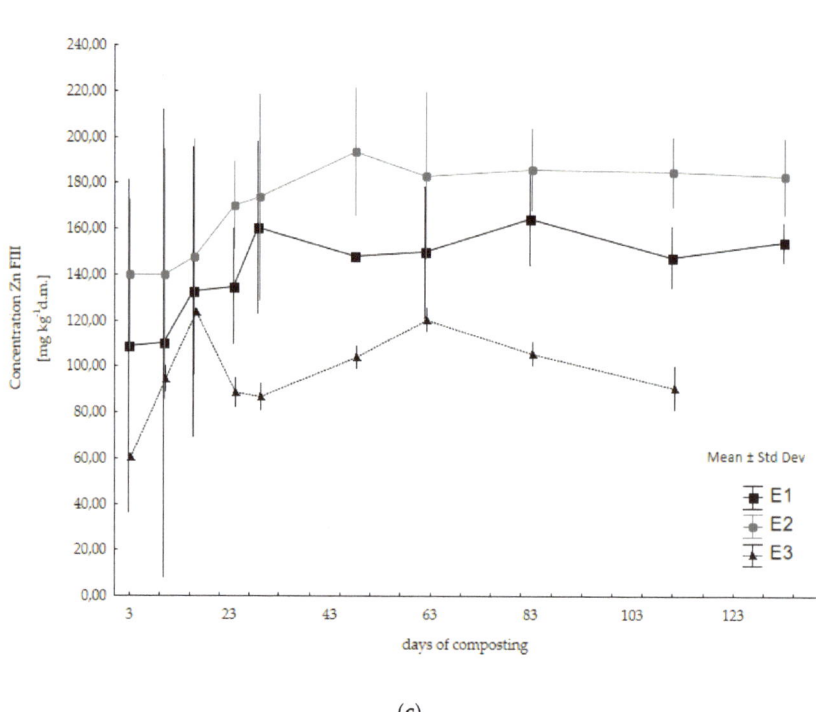

(c)

Figure 3. *Cont.*

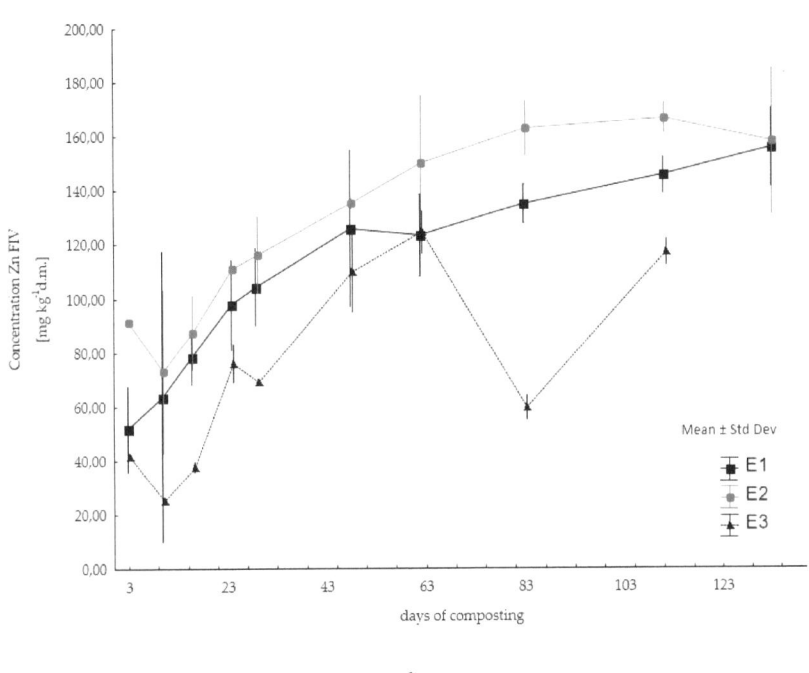

(d)

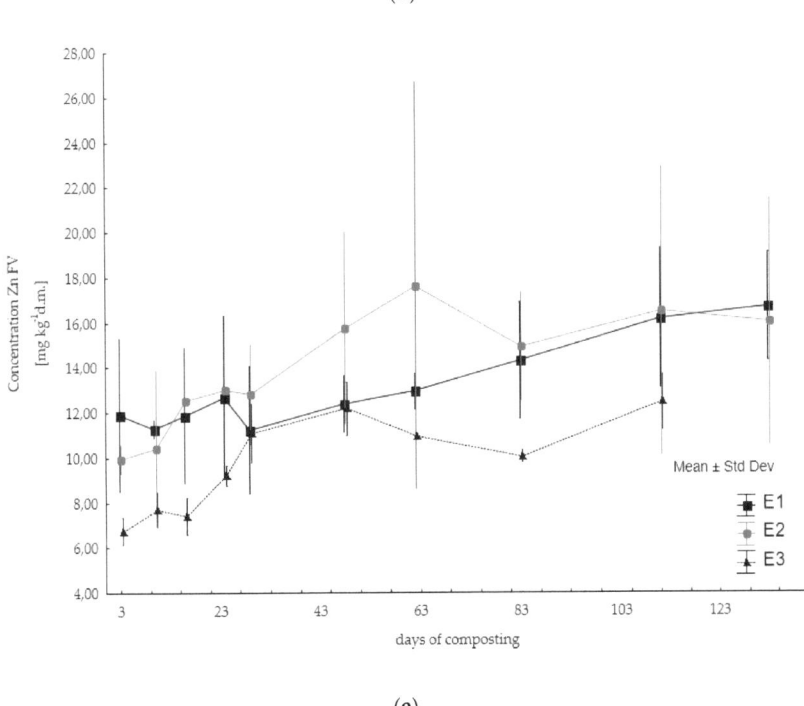

(e)

Figure 3. Concentration of Zn in particular fractions FI (**a**), FII (**b**), FIII (**c**), FIV (**d**), and FV (**e**) in stages E1, E2, E3.

During composting, at all stages, reduction of Zn compounds' contents in the most mobile fractions, i.e., ion exchangeable (FI) and carbonate (FII), occurred. An increase of organic matter-bound zinc forms' contents was noted. In samples taken from windrows at the E1 and E2 stages, an increase of zinc compounds with Fe/Mn oxides' contents as well as a decrease of their percentage in the residual fraction were noted during composting. In the case of compost samples taken at the E3 stage, the trend was reversed (Table 5). The highest percentage in FV was noted for samples taken at the E2 stage, making up from 15.37% of the total content at the beginning of the process down to 9.79% on the last day. The highest increase of Zn compounds in organic fractions was noted for E3 (16.15) and E1 (15.93). In samples taken from the windrow on the last day of the process, the percentage of ion exchangeable Zn compounds was the lowest at stage 3 and amounted to 1.32% and the highest was at E1, at 2.72%. The lowest percentage in the carbonate fraction during the process course was at E3. Samples taken during the last process day at E2 manifested the highest Zn compounds' percentage in fraction III, at 37.33%, whereas at E3 the share of Zn compounds bound with Fe/Mn oxides was the lowest and amounted to 30.80%. The highest content of Zn compounds (last day) bound with organic substance was noted for E3, at 39.68%, and the lowest for E2, at 31.69%.

During composting the percentage of Zn compounds in the ion exchangeable (FI) and carbonate (FII) fractions decreased. The highest loss of ion exchangeable (FI) Zn compounds was noted in the compost of initial C/N = 9.20 (E3). Compost samples taken at E1 and E2 featured a higher loss of those Zn forms contents, which were bound with carbonates, compared with E3.

3.1.3. Cu and Zn Mobility Factor

The MF mobility factor, calculated as the ratio of the heavy metal sum in fractions F1 and FII to the total content, may be used to asses potential mobility of heavy metals [14]. During the composting process, MF values for tested elements decreased at all composting stages, which proves a reduction of Cu and Zn mobility. Similar results were obtained by Wu et al. and Wang et al. [14,18]. The MF value for Zn was much higher than that for Cu, thus indicating that potential Zn mobility was higher during composting than that of Cu. The lowest MF values for Cu, during the composting process, were noted for E3 compost (C/N = 9.20) and amounted from 4.99 to 1.18. This compost featured the lowest MF value decrease during the composting process. The highest MF value decrease was noted for compost in which C/N amounted to 9.61—E2 (decrease by 11.51), although at this stage Cu content in the FI and FII fractions was highest. The MF values for samples taken at the end of the composting process at the E1, E2, and E3 stages amounted to 2.95, 2.92, and 2.39, respectively.

The Zn MF value for all stages remained within the 38.40 (initial value) to 17.42 (final value) interval. The lowest Zn MF values were noted for those compost samples that featured the lowest initial C/N value (E3, 9.20). At all composting stages, the MF value decreased by approximately 17%. In those composts that had initial C/N values of 14.56 (E1), the MF value by the end of the composting process was highest (Table 5).

Reduced Cu MF may protect plants against Cu in soil with compost added, whereas higher Zn bioavailability promotes alleviation of Zn deficit in the human food chain.

Metals that occur in the organic matter (FIV) and residual (FV) bound fractions show low mobility and minor bioavailability. Inactivation rates (*IR*) for Cu and Zn were calculated in order to assess the change of exchangeable fraction content percentage before and after the composting process. Cu *IR* values for E1, E2, and E3 amounted to 77.7, 80.30, and 61.3, respectively (Table 5). The highest impact on Cu immobilization was noted during the composting process of biomass with initial C/N = 9.61 (E2). Zn *IR* values were highest in compost samples with initial C/N = 9.61 (E2) and 9.20 (E3), which amounted to 50.3 and 51.9, respectively. High *IR* values prove a significant impact of the composting process on copper and zinc passivation. Xu et al. [13] obtained in their research work other results and proved the absence of composting impact on Cu and Zn transformation into forms that are hardly available for living organisms.

3.2. Numerical Analysis

An artificial neural network of classification type was used to perform numerical analysis. Using test results, a database composed of 32 cases, 13 independent variables and one dependent variable, was created. As the proportion between the cases and independent variables (predictors) was about 3:1 and was lower than the recommended 10:1 [40], the number of predictors was reduced to three, i.e., time, C/N, and total content of the given element. The dependent variable was a qualitative variable marked with letters A and B referring to compost quality classes in accordance with the regulations applicable in Austria [12]. The criterion of qualification of each case to a given class was the value of the resultant variables X(Cu) and Y(Zn), making a sum of concentrations of a given element determined in all four fractions except for fraction V. The classification was performed based on the following rule:

If the concentration was $150 < X(Cu) \leq 500$ mg·kg^{-1}d.m., then class B;
If the concentration was $70 < X(Cu) \leq 150$ mg·kg^{-1}d.m., then class A;
If the concentration was $500 < Y(Zn) \leq 1800$ mg·kg^{-1}d.m., then class B;
If the concentration was $200 < Y(Zn) \leq 500$ mg·kg^{-1}d.m., then class A;
If the concentration was $Y(Zn) \leq 200$ mg·kg^{-1}d.m., then class A+

Due to the simulations being performed separately for the two analyzed elements, i.e., Cu and Zn, neural networks of the architectures MLP 3-5-3 and MLP 3-6-3 were chosen, respectively. Correctness of the dependent variable prediction results for the cases used at the network teaching stage is shown in Table 6. Correctness of estimates at the network testing stage amounted to 100% in both tested heavy metal cases.

Table 6. Prediction of correctness for the teaching set.

Dependent Variable		Number of Estimates			
		Category			All
		a+	a	b	
Cu	correct	0	5	12	17
	incorrect	2	4	1	7
Zn	correct	0	23	0	23
	incorrect	1	0	0	1

The method of learning that was aimed at minimizing the neural network error values following modification of values of the weight coefficients of neuron input signals was the Quasi-Newton (BFGS) algorithm [41].

The analysis of sensitivity defining the weight of the independent variables (Table 7), as predictors in the adopted NN model, proved significance of all three variables, i.e., time, C/N, and total given element content for the neural network prediction quality.

Table 7. Independent variables sensitivity test results.

NN	Independent Variables			Quality of	
	Total Cu, Zn	C/N	Time	Teaching	Testing
Cu—MLP 3-5-3	1.015	1.413	1.008	70.83	100
Zn—MLP 3-6-3	1.004	1.005	0.999	95.83	100

The developed model of the artificial neural network allows to determine the real risk of compost contamination with heavy metals. Some heavy metals can be permanently bound in the soil matrix-fraction no. V. Desorption of heavy metals bound by, e.g., clay minerals, theoretically occurs at pH = 1 but in real environmental conditions is unlikely [15,17,18]. The analysis of the values of the parameters constituting independent variables in the proposed model, i.e., the total content of a given element and the concentration of organic carbon and total nitrogen at any time during the composting process, allows for compost classification taking into account the presence of only mobile fractions, and is thus potentially hazardous to the environment.

4. Conclusions

Although sewage sludge is a source of fertilizers, unfortunately, it contains heavy metal concentrations that can exceed admissible values. In our research work we analyzed the trend of Cu ad Zn chemical forms' transformations under reduced supplementation during composting conditions, taking into account:

(1) The impact of limited volumes of straw added as a source of organic carbon;
(2) The presence of mobile forms of tested metals based on sequential extraction;
(3) The classification of mature compost based on adopted standards using a neural network of standard classification type.

Our research work proved that composting of a sewage sludge mix with relatively low structural material input and, consequently, low initial C/N values, had no negative impact on biochemical transformations' velocity. The initial C/N value had an impact on the distribution of both tested elements in a manner correlated with the partial decomposition of organic substance. A lower risk of Cu and Zn liberation due to reduction of mobile heavy metals' fractions' shares with a simultaneous increase of their share in stable fractions was ascertained.

The developed neural network is a tool allowing to predict compost classes depending on three parameters, including C/N. The C/N value depends on the share of the supplement making an additional source of organic carbon, added at the batch formation stage. Therefore, using the developed NN model, the simulation of compost quality in accordance with classifications applicable in Austria for various initial conditions and any composting time can be performed.

Author Contributions: Data curation, A.L.; Methodology, B.J.; Software, R.S. All authors have read and agreed to the published version of the manuscript.

Funding: This research was partly funded by the European Union Interreg South Baltic Program, "Sludge Technological Ecological Progress-increasing the quality and reuse of sewage sludge" number STHB.02.02.00-32-0110/17.

Institutional Review Board Statement: Not applicable.

Informed Consent Statement: Not applicable.

Data Availability Statement: Not applicable.

Conflicts of Interest: The authors declare no conflict of interest.

References

1. Sidełko, R.; Siebielska, I.; Janowska, B.; Skubała, A. Assessment of biological stability of organic waste processed under aerobic conditions. *J. Clean. Prod.* **2017**, *164*, 1563–1570. [CrossRef]
2. GUS—Statistics Poland 2020. Environment Statistics Reports for the Poland. Available online: https://stat.gov.pl/en/ (accessed on 24 September 2020).
3. Kujawa, S.; Mazurkiewicz, J.; Czekała, W. Using convolutional neural networks to classify the maturity of compost based on sewage sludge and rapeseed straw. *J. Clean. Prod.* **2020**, *258*, 120814. [CrossRef]
4. Carrizo, M.E.; Alesso, C.A.; Cosentino, D.; Imhoff, S. Aggregation agents and structural stability in soils with different texture and organic carbon content. *Sci. Agric.* **2015**, *72*, 75–82. [CrossRef]
5. Curtis, M.J.; Claassen, V.P. Regenerating topsoil functionality in four drastically disturbed soil types by compost incorporation. *Restor. Ecol.* **2009**, *17*, 24–32. [CrossRef]
6. Pinasseau, A.; Zerger, B.; Roth, J.; Canova, M.; Roudier, S. *Best Available Techniques (BAT) Reference Document for Waste Treatment Industrial Emissions Directive 2010/75/EU (Integrated Pollution Prevention and Control)*, EUR 29362 EN; Publications Office of the

7. European Union: Luxembourg, 2018; p. JRC113018. Available online: https://publications.jrc.ec.europa.eu/repository/handle/JRC113018 (accessed on 22 October 2018).
7. Publications Office of the European Union. Commission Decision. Establishing Ecological Criteria for the Award of the Community Eco-Label to Soil Improvers and Growing Media. 2001/688/EC. 2001. Available online: https://op.europa.eu/en/publication-detail/-/publication/ (accessed on 20 May 2005).
8. Publications Office of the European Union. *Council Directive 86/278/EEC of 12 June 1986 on the Protection of the Environment, and in Particular of the Soil, When Sewage Sludge Is Used in Agriculture*; European Commission: Brussels, Belgium, 1986. Available online: https://op.europa.eu/en/publication-detail/-/publication/f76faa39-2b27-42f2-be1e-9332f795e324/language-en/ (accessed on 12 June 1986).
9. Singh, J.; Kalamdhad, A.S. Assessment of bioavailability and leachability of heavy metals during rotary drum composting of green waste (Water hyacinth). *Ecol. Eng.* **2013**, *52*, 59–69. [CrossRef]
10. Zhou, H.; Meng, H.; Zhao, L.; Shen, Y.; Hou, Y.; Cheng, H.; Song, L. Effect of biochar and humic acid on the copper, lead, and cadmium passivation during composting. *Bioresour. Technol.* **2018**, *258*, 279–286. [CrossRef]
11. Regulation (EU) 2019/1009 of the European Parliament and of the Council of 5 June 2019 Laying Down Rules on the Making Available on the Market of EU Fertilising Products and Amending Regulations (EC) No 1069/2009 and (EC) No 1107/2009 and Repealing Regulation (EC) No 2003/2003. Available online: https://eur-lex.europa.eu/legal-content/EN/TXT/HTML/?uri=CELEX:32019R1009&from=EN (accessed on 25 June 2019).
12. Biala, J.; Wilkinson, K. International Comparison of the Australian Standard for Composts, Soil Conditioners and Mulches. University of Queensland, Australia. 2020. Available online: https://www.aora.org.au/sites/default/files/uploaded-content/website-content/International_Comparison_AS4454_Final.pdf (accessed on 15 February 2020).
13. Xu, S.; Li, L.; Zhan, J.; Guo, X. Variation and factors on heavy metal speciation during co-composting of rural sewage sludge and typical rural organic solid waste. *J. Environ. Manag.* **2022**, *306*, 114418. [CrossRef]
14. Wu, S.; Shen, Z.; Yang, C.; Zhou, Y.; Li, X.; Zeng, G.; Ai, S.; He, H. Effects of C/N ratio and bulking agent on speciation of Zn and Cu and enzymatic activity during pig manure composting. *Int. Biodeterior. Biodegrad.* **2017**, *119*, 429–436. [CrossRef]
15. Mossop, K.F.; Davidson, C.M. Comparison of original and modified BCR sequential extraction procedures for the fractionation of copper, iron, lead, manganese and zinc in soils and sediments. *Anal. Chim. Acta* **2003**, *478*, 111–118. [CrossRef]
16. Gondek, K.; Mierzwa-Hersztek, M.; Kopec, M. Mobility of heavy metals in sandy soil after application of composts produced from maize straw, sewage sludge and biochar. *J. Environ. Manag.* **2018**, *210*, 87–95. [CrossRef]
17. He, M.-M.; Tian, G.-M.; Liang, X.-Q. Phytotoxicity and speciation of copper, zinc and lead during the aerobic composting of sewage sludge. *J. Hazard. Mater.* **2009**, *163*, 671–677. [CrossRef] [PubMed]
18. Wang, L.; Li, Y.; Prasher, S.O.; Yan, B.; Ou, Y.; Cui, H.; Cui, Y. Organic matter, a critical factor to immobilize phosphorus, copper, and zinc during composting under various initial C/N ratios. *Bioresour. Technol.* **2019**, *289*, 121745. [CrossRef] [PubMed]
19. Wang, X.; Chen, T.; Zheng, G. Preservation of nitrogen and sulfur and passivation of heavy metals during sewage sludge composting with KH_2PO_4 and $FeSO_4$. *Bioresour. Technol.* **2020**, *297*, 122383. [CrossRef]
20. Bernal, M.P.; Alburquerque, J.A.; Mora, R. Composting of animal manures and chemical criteria for compost maturity assessment. A review. *Bioresour. Technol.* **2009**, *100*, 5444–5453. [CrossRef] [PubMed]
21. Sidełko, R.; Janowska, B.; Walendzik, B.; Siebielska, I. Two composting phases running in different process conditions timing relationship. *Bioresour. Technol.* **2010**, *101*, 6692–6698. [CrossRef]
22. Kacprzak, K.; Neczaj, E.; Fijałkowski, K.; Grobelaka, A.; Grosser, A.; Worwag, M.; Rorat, A.; Brattebo, H.; Almas, A.; Singh, B.R. Sewage sludge disposal strategies for sustainable development. *Environ. Res.* **2017**, *156*, 39–46. [CrossRef]
23. Świerczek, L.; Cieślik, B.M.; Konieczka, P. The potential of raw sewage sludge in construction industry—A review. *J. Clean. Prod.* **2018**, *200*, 342–356. [CrossRef]
24. Doublet, J.; Francou, F.; Poitrenaud, M.; Houot, S. Sewage sludge composting: Influence of initial mixtures on organic matter evolution and N availability in the final composts. *Waste Manag.* **2010**, *30*, 1922–1930. [CrossRef]
25. Zhau, Y.; Selvam, A.; Wong, J.W.C. Chinese medicinal herbal residues as a bulking agent for food waste composting. *Bioresour. Technol.* **2018**, *249*, 182–188. [CrossRef]
26. Graca, J.; Murphy, B.; Pentlavalli, P.; Allen, C.C.R.; Bird, E.; Gaffney, M.; Duggan, T.; Kelleh, B. Bacterium consortium drives compost stability and degradation of organic contaminants in in-vessel composting process of the mechanically separated organic fraction of municipal solid waste (MS-OFMSW). *Bioresour. Technol.* **2021**, *13*, 100621.
27. Gonzalez, D.; Colon, J.; Gabriel, D.; Sanchez, A. The effect of the composting time on the gaseous emissions and the compost stability in a full-scale sewage sludge composting plant. *Sci. Total Environ.* **2019**, *654*, 311–323. [CrossRef] [PubMed]
28. Zheng, G.; Wang, T.; Niu, M.; Chen, X.; Liu, C.; Wang, Y.; Chen, T. Biodegradation of nonylphenol during aerobic composting of sewage sludge under two intermittent aeration treatments in a full-scale plant. *Environ. Pollut.* **2018**, *238*, 783–791. [CrossRef] [PubMed]
29. Głąb, T.; Żabiński, A.; Sadowska, U.; Gondek, K.; Kopeć, M.; Mierzwa-Hersztek, M.; Taborc, S. Effects of co-composted maize, sewage sludge, and biochar mixtures on hydrological and physical qualities of sandy soil. *Geoderma* **2018**, *315*, 7–35. [CrossRef]
30. Li, S.; Li, D.; Li, J.; Li, G.; Zhang, B. Evaluation of humic substances during co-composting of sewage sludge and corn stalk under different aeration rates. *Bioresour. Technol.* **2017**, *245*, 1299–1302. [CrossRef] [PubMed]

31. Cerda, A.; Artola, A.; Font, X.; Barrena, R.; Gea, T.; Sanchez, A. Composting of food wastes: Status and challenges. *Bioresour. Technol.* **2018**, *248*, 57–67. [CrossRef]
32. Kulikowska, D.; Sindrewicz, S. Effect of barley straw and coniferous bark on humification process during sewage sludge composting. *Waste Manag.* **2018**, *79*, 207–213. [CrossRef]
33. Sidełko, R.; Walendzik, B.; Smuga-Kogut, M.; Janowska, B.; Szymański, K.; Głowacka, A.; Leśniańska, A. Impact of reduced straw content on the sewage sludge composting process. *Arch. Environ. Prot.* **2020**, *46*, 70–77.
34. Szymański, K.; Janowska, B.; Sidełko, R. Estimation of bioavailability of copper, lead and zinc in municipal solid waste and compost. *Asian J. Chem.* **2005**, *17*, 1646–1660.
35. Interreg South Baltic 2018. STEP. Sludge Technological Ecological Progress—Increasing the Quality and Reuse of Sewage Sludge. Project No. STHB.02.02.00-32-0110/17. Available online: https://www.step-interreg.eu/pl/ (accessed on 28 October 2020).
36. *PN-R-04006: 2000*; Organic Fertilizers—Collection and Preparation of Manure and Compost Samples. PKN: Warsaw, Poland, 2000.
37. *PN-Z-15011-3: 2001*; Municipal Solid Waste Compost—Determination of pH, Content of Organic Substance, Organic Carbon, Nitrogen, Phosphorus and Potassium. PKN: Warsaw, Poland, 2001.
38. Zhu, N.-M.; Guo, X.-J. Sequential extraction of anaerobic digestate sludge for the determination of partitioning of heavy metals. *Ecotoxicol. Environ. Saf.* **2014**, *102*, 18–24. [CrossRef]
39. Kang, J.; Zhang, Z.; Wang, J.J. Influence of humic substances on bioavailability of Cu and Zn during sewage sludge composting. *Bioresour. Technol.* **2011**, *102*, 8022–8026. [CrossRef]
40. StatSoft Electronic Statistic Textbook PL 2006. Available online: http://www.statsoft.pl/textbook/stathome.html (accessed on 30 September 2006).
41. Gaffke, N.; Schwabe, R. Quasi-Newton algorithm for optimal approximate linear regression design: Optimization in matrix space. *J. Stat. Plan. Inference* **2019**, *198*, 62–78. [CrossRef]

Article

Effect of Filter Medium on Water Quality during Passive Biofilter Activation in a Recirculating Aquaculture System for *Oncorhynchus mykiss*

Arkadiusz Nędzarek [1,*], Małgorzata Bonisławska [1], Agnieszka Tórz [1], Adam Tański [2] and Krzysztof Formicki [2]

[1] Department of Aquatic Bioengineering and Aquaculture, Faculty of Food Sciences and Fisheries, West Pomeranian University of Technology in Szczecin, Kazimierza Królewicza 4, 71-550 Szczecin, Poland
[2] Department of Hydrobiology, Ichthyology and Biotechnology of Reproduction, Faculty of Food Sciences and Fisheries, West Pomeranian University of Technology in Szczecin, Kazimierza Królewicza 4, 71-550 Szczecin, Poland
* Correspondence: anedzarek@zut.edu.pl

Highlights:

What are the main findings?

- The high specific surface area of the biofilter media has a positive effect on nitrification.
- The smooth surface of the biofilter medium reduces the efficiency of nitrification.

What is the implication of the main finding?

- Passive biofilter activation can be used in salmonid RAS.
- P and C concentrations do not limit nitrogen transformation processes.

Abstract: High-performance biofilters for water purification in recirculating aquaculture systems (RAS) ensure the safety of cultures of highly nutritious fish. As the most critical step in the functioning of biofilters is their activation, the objective of this study was to evaluate the suitability of commercial artificial media, namely RK Plast (BR-1), Mutag-BioChip30 (BR-2), and LevaPor (BR-3), for the passive activation of biofilters used in rainbow trout farming. Changes in NH_4^+-N, NO_2^--N, NO_3^--N, phosphorus, and carbon concentrations were analyzed. In the first period, an increase in NH_4^+-N concentration was recorded, before an increase in NO_2^--N concentration (maximum concentrations ranged 0.728–1.290 and 0.982–5.198 mg N dm^{-3}, respectively), followed by a reduction and stabilization to a level safe for the fish (both below 0.100 mg N dm^{-3}). Concurrently, a steady increase in NO_3^--N concentration was noted, with a maximum concentration between 6.521 and 7.326 mg N dm^{-3}. Total phosphorus and total carbon ranged from 0.423 to 0.548 mg P dm^{-3}, and from 43.8 to 45.2 mg C dm^{-3}. The study confirmed the feasibility of using the tested artificial biofilter media for rainbow trout farming in RAS with passive biofilter activation. Biofilter activation efficiency was highest for the media with the highest specific surface area (BR-2 and BR-3). The removal of ammonium nitrogen and nitrite nitrogen was above 90%. Nitrogen biotransformation was not limited by phosphorus or carbon concentrations.

Keywords: salmonid fish; rainbow trout; RAS; nitrogen; phosphorus; carbon

1. Introduction

Fish is a very significant food source, with an annual growth in consumption of about 2.1%. In 2018, 82 million tonnes of fish were produced from aquaculture, accounting for almost 50% of global production [1].

The over-exploitation of natural fish reserves and the ever-increasing environmental restrictions on traditional aquaculture are favoring the development of intensive aquaculture in recirculation systems (RAS), where the negative impact on the environment is much lower than in open systems. At the same time, the introduction of technological innovations and the rising standards of fish farming in RAS are an opportunity to obtain fish products that are safe for consumer health. Modern recirculating aquaculture integrates a number of devices in an automated system, such as microfilters, smart feeding devices, and water degassing, quality treatment and monitoring systems, as well as biofilters necessary for biological water purification [2–4].

Biofilters convert ammonia nitrogen to nitrite nitrogen and then to the less toxic nitrate nitrogen for the fish [5,6]. This process, known as nitrification, is carried out by biofilm-forming bacteria on the biofilter medium. The efficiency of the biofilter is related to the surface area that the bacteria colonize. The greater the surface area of the medium used, the greater the area for colonization by bacteria and contact with the medium, resulting in a biofilter with more bacteria removing ammonia from the system [7,8].

Bacterial substrates are non-corroding materials, such as fiberglass, ceramic, rock, or plastic, with increasingly popular plastic molds producing a variety of shapes with a high surface area per unit of volume (usually referred to as the specific surface area—SSA). However, biofiltration media with a higher SSA can be more easily clogged by the bacteria (biofouling) than media with lower SSA, resulting in a reduced biofilter performance. There must, therefore, be a balance between a high SSA and an operationally reliable biofilter [6,8].

One important step in the operation of a water treatment system in a RAS, is the activation of the biofilter, with one strategy involving passive activation (the cold start method) in which the fish culture is conducted without a previously activated biofilter. This method has the advantage of using bacteria that are introduced into the culture system along with the fish. However, due to the potential for high increases in ammonia and nitrite concentrations, it requires a higher water exchange rate and a reduction in feed intake until the biofilter is activated [9,10].

It should also be borne in mind that seed sludge for the inoculation of bioreactors in RAS is not always available, and that seed sludge from municipal or industrial wastewater treatment systems is not recommended because it may contain pathogens. Thus, the method of passive activation, despite some limitations, may be the only one that can be used, especially for breeding fish with high environmental requirements, for euryhaline species (such as salmon), and also when it is desirable to protect the culture from the influence of foreign microflora. Passive activation can reduce stress not only for the fish, but also the microorganisms already inhabiting the bioreactors [10,11].

The aim of this study then was to evaluate the passive activation of selected biofilter beds in a recirculating aquaculture system used for culturing cold-water salmonid fish (rainbow trout, *Oncorhynchus mykiss*), and to compare changes in the concentration of nitrogen, phosphorus, and carbon during this process, between the shapes and specific surface areas of the various plastic substrates (non-porous medium with a smooth surface, hard porous medium and soft porous medium with activated carbon). The experiments were performed as part of a pilot study to select the most suitable bed for the water treatment system in the *O. mykiss* breeding facility under construction.

2. Materials and Methods

Laboratory-scale recirculating systems for rainbow trout culturing were used to study the process of biofertilizer passive activation. The systems were located in the isothermal laboratory of the Faculty of Food Sciences and Fisheries, West Pomeranian University of Technology in Szczecin (Poland). The efficiencies of the biochemical processes in the biofilters were evaluated by analyzing the variability of particular physicochemical parameters of the water circulating in the culture systems.

2.1. Recirculating System

Each recirculating system consisted of a circular fish rearing tank (capacity 1 m^3) filled with distilled tap water to a volume of 0.54 m^3, and two bioreactors with a test medium (Figure 1). The bioreactor was a FLUVAL FX-6 from Rolf C. Hagen Corp. (Mansfield, MA,

USA) canister filter with a total volume of 0.02 m³. The volume of the media baskets in the bioreactor was 0.0059 m³. The water flow rate through the bioreactor was 0.00055 m³ s⁻¹. The bioreactor outflows were set so that the water in the fish pool circulated clockwise, forcing the fish to position themselves in the water current. Ten percent of the water was replaced with fresh water each day. The water in the culture tanks was continuously oxygenated using atmospheric air dispersion, maintaining a dissolved oxygen concentration between 90–98%. Water temperature was maintained at 12 ± 2 °C and water pH was in the range of 7–8. The rearing was carried out in a day/night 12/12 h day/night light cycle.

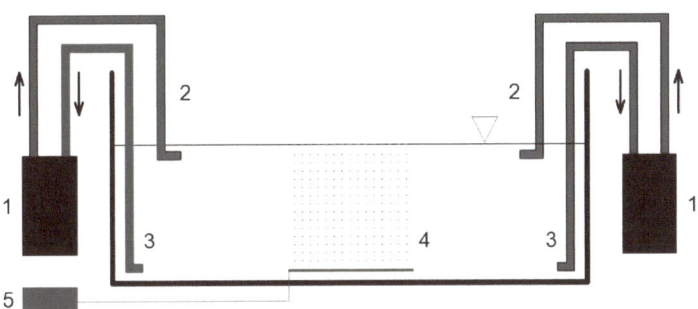

Figure 1. Diagram of the experimental culture system: 1—bioreactor; 2—bioreactor water outlet; 3—water intake from the culture tank; 4—atmospheric air diffuser; 5—air pump.

Twenty fish, with an initial average length of 235 ± 1 mm and an average weight of 168.7 ± 0.5 g, were reared in each tank. The fish were fed 4 times a day with a total of 50 g of Aller Gold pellets from Aller Aqua (feed for rainbow trout fattening).

The following commercial media were assessed for filtration: RK-Plast (BR-1), Mutag-BioChip30 (BR-2), and LevaPor (BR-3). The general characteristics of the media are shown in Table 1.

Table 1. Comparison of selected commercial bioreactor media in the recirculating aquaculture systems.

		RK Plast	Mutag-BioChip30	LevaPor
		BR-1	BR-2	BR-3
Shape		saddle	round chips	cube
Size	mm	30 × 15	30 × 1.1	20 × 20 × 7
Weight	kg m⁻³	145	165	26–28
Surface (SSA)	m² m⁻³	700	5500	2700
Composition		Polypropylene	Polyethylene	Polyurethane + activated carbon

2.2. Hydrochemical Parameters

Water samples were collected once a week for 14 weeks from the outflow and inlet of the bioreactor. Nitrite nitrogen (NO_2^--N), nitrate nitrogen (NO_3^--N), ammonium nitrogen (NH_4^+-N), total nitrogen (TN), total reactive phosphorus (TRP), total phosphorus (TP), total inorganic carbon (TIC), and total organic carbon (TOC) were determined in the sampled water. Chemical analyses were performed according to methodologies recommended by APHA [12]. A HITACHI (Tokyo, Japan) UV-VIS U-2900 spectrophotometer was used for colorimetric analyses. Carbon and TN determinations were performed in infrared using a VarioTOC SELECT from ELEMENTAR (Langenselbold, Germany).

2.3. Statistical Analysis

The results obtained from this study were analyzed using one-way ANOVA, Tukey's post hoc tests, using Statistica v13.3 software from TIBCO Software Inc. (Palo Alto, CA, USA). We evaluated the significance of differences in the concentration of the determined hydrochemical indicators (i) between the water flowing into and out of the bioreactors, (ii) and differences between the tested bioreactor media. The differences were considered significant at $p < 0.05$. Linear regressions were also determined for variation in the inorganic forms of nitrogen over a function of time.

3. Results and Discussion

Comparing the concentrations of nitrogen, phosphorus, and carbon in the water flowing into the bioreactors with their concentrations in the water flowing out of the bioreactors revealed no significant differences ($p > 0.05$) (see Appendix A Table A1). This effect is explained by the short water retention time in the bioreactors, which is necessary to ensure an optimal level of oxygenation in the biofilter zone, and which is essential for efficient nitrogen biotransformation. For example, Dias et al. [13] showed that the rate of nitrification increases with an increase in the flow capacity of the recirculation system, as this promotes water circulation and oxygen distribution in all areas of the media.

Following this, the paper then focused on: (i) changes in the concentration of N, P, and C as a function of the duration of the experiment, and (ii) differences in the concentration of these indicators between the tested media.

3.1. Changes in the Concentrations of N, P, and C during the Experiment

Generally, as shown in Table 2, the indicators determined had a high range of concentration variability (wider for nitrogen, and narrower for phosphorus and carbon). For example, the concentration ranges (data for BR-1; in mg N, P, or C dm^{-3}) ranged from 0.009 to 5.198 for NO_2^--N, from 0.072 to 0.533 for TRP, and from 23.7 to 45.2 for TIC.

Three distinct trends of changes were observed. (1) For ammonium nitrogen and nitrite nitrogen, the concentrations first increased to a maximum value (for ammonium nitrogen, 1.290, 0.834, and 0.728 mg N dm^{-3}; and for nitrite nitrogen, 5.198, 4.382, and 0.982 mg N dm^{-3}; for BR-1, BR-2, and BR-3, respectively), and then decreased to a relatively stable level by the end of the experiment. This stabilization occurred at 8 weeks for BR-1, at 5 weeks for BR-2, and at 4 weeks for BR-3. At the respective times, the mean concentrations (in mg N dm^{-3}) of ammonium nitrogen were 0.071, 0.068, and 0.056, and nitrite nitrogen at 0.092, 0.079, and 0.079 for BR-1, BR-2, and BR-3, respectively. The reduction of ammonium nitrogen from maximum values to steady-state concentrations was 94.5% (BR-1), 91.8% (BR-2), and 92.3% (BR-3). The respective reductions in nitrite nitrogen were 98.2% (BR-1 and BR-2), and 91.9% (BR-3). (2) For NO_3^--N, TN, TRP, and TP, the observed trends were characterized by a slow increase in concentration, reaching maximum values at the end of the experiment. (3) For TIC and TOC, the maximum values were shown during the first weeks of the experiment, followed by a decrease in concentration and a relatively sustained stabilization for the following weeks. A reduction in carbon concentration in the systems occurred after 6 weeks for both BR-1 and BR-2, and after 4 weeks for BR-3, to about 26–28 mg C dm^{-3} for TIC, and about 10 mg C dm^{-3} for TOC (Appendix A Figure A1).

Table 2. Concentrations of the studied forms of nitrogen, phosphorus, and in the water of recirculating systems with bioreactors with selected filtration media: RK Plast (BR-1), Mutag-BioChip30 (BR-2), and LevaPor (BR-3).

			Bioreactor		
			BR-1	BR-2	BR-3
NO_2^--N	mg N dm^{-3}	range	0.009–5.198	0.012–4.382	0.016–0.982
		mean	0.828 [b]	0.439 [ab]	0.165 [a]
		median	0.128	0.086	0.076
		SD	1.382	1.081	0.246
NO_3^--N	mg N dm^{-3}	range	0.198–6.521	0.335–6.778	0.600–7.326
		mean	2.851 [a]	3.585 [a]	3.496 [a]
		median	2.754	3.985	3.774
		SD	2.030	1.887	1.908
NH_4^+-N	mg N dm^{-3}	range	0.024–1.290	0.026–0.834	0.020–0.728
		mean	0.271 [b]	0.158 [ab]	0.105 [a]
		median	0.084	0.075	0.048
		SD	0.350	0.206	0.168
TN	mg N dm^{-3}	range	1.230–12.749	1.302–12.716	1.067–15.859
		mean	7.790 [a]	8.310 [a]	9.711 [a]
		median	8.367	8.562	10.617
		SD	3.523	3.456	4.182
TRP	mg P dm^{-3}	range	0.072–0.533	0.092–0.412	0.075–0.529
		mean	0.187 [a]	0.188 [a]	0.293 [b]
		median	0.214	0.150	0.239
		SD	0.127	0.096	0.1560
TP	mg P dm^{-3}	range	0.102–0.548	0.119–0.423	0.132–0.541
		mean	0.251 [a]	0.242 [a]	0.326 [b]
		median	0.232	0.229	0.272
		SD	0.116	0.077	0.152
TIC	mg C dm^{-3}	range	23.7–45.2	23.9–44.4	19.3–43.8
		mean	32.3 [a]	31.9 [a]	29.6 [a]
		median	30.0	28.9	27.6
		SD	6.0	6.3	6.5
TOC	mg C dm^{-3}	range	9.5–14.6	9.2–18.8	8.6–14.1
		mean	11.1 [b]	11.6 [b]	10.3 [a]
		median	11.0	11.0	9.6
		SD	1.2	2.1	1.6

[a,b] different symbols in rows indicate significant differences between the tested bioreactor media (ANOVA, Tukey HSD test, $p < 0.05$).

The observed trends of changes in the concentration of the noted forms of nitrogen, phosphorus, and carbon are characteristic of the biofilter activation stage in RAS systems [4]. The sources of these elements are fish metabolic products and uneaten feed [14]. For example, Avnimelech and Ritvo [15] report that the average intake of organic carbon, nitrogen, and phosphorus from the feed by fish are approximately 13%, 29%, and 16%, respectively. With unrestricted access to carbon and phosphorus, nitrogen compounds can be efficiently transformed by microorganisms and used for biomass production [16]. At the same time, during the biofilter activation phase, an increase in ammonium nitrogen concentration is observed in the first phase, preceding the increase in nitrite nitrogen concentration; then, the concentrations decrease to a stable low level that was safe for fish. The time of the first phase of this process can last about 2 weeks [14]; however, it can be prolonged at lower temperatures and in passive activation [9,10,17], similar to our study.

It should also be borne in mind that in RAS systems, the wide range of N and P concentrations is associated with the fish species among other things, and, thus, on the culture conditions needed to breed the species (e.g., water temperature, feed chemistry) and the operating conditions of the RAS system (e.g., biofilter volume, water retention time). As an example, in the first weeks of culture using experimental RAS systems for rearing goldfish (*Carassius auratus auratus*) and koi (*Cyprinus carpio koi*), Sikora et al. [18] recorded nitrite nitrogen concentrations of 7.3 and 10.7 mg N dm^{-3}, respectively, and ammonia nitrogen at >27 mg N dm^{-3}. Even higher concentrations of ammonia nitrogen (>35 mg N dm^{-3}) were recorded by Owatari et al. [19] in a Nile tilapia (*Oreochromis niloticus*) culture system. On the other hand, Żarski et al. [20] recorded low concentrations of ammonia nitrogen (0.5 mg N dm^{-3}) and nitrite nitrogen (0.3 mg N dm^{-3}) in an experiment with juvenile stages of ide (*Leuciscus idus*). In our experiment, ammonia nitrogen concentrations in the first weeks of culturing were consistent with other studies on RAS systems for *O. mykiss*, as in the study by Pulkkinen et al. [21], who reported about 1.2 mg N dm^{-3}. At the same time, the nitrite nitrogen concentrations in our study were at a higher level than in the study of Pulkkinen et al. [21] (about 1.0 mg dm^{-3}). The differences may have been the result of the more intensive biochemical transformation of nitrogen compounds conditioned by the higher water temperature in their study (16 °C). Moreover, the concentrations of ammonium nitrogen and nitrite nitrogen in the steady state recorded by us were consistent, for example, with the results of Fernandes et al. [22] (0.12 and 0.26 mg N dm^{-3}, respectively), and also with the results of Pulkkinen et al. [21,23], who emphasized that the biofilters they tested were effective and produced circulating water that met the requirements of cultured *O. mykiss*. The reductions in ammonium nitrogen and nitrite nitrogen obtained in our study were consistent with the results of other authors, who also obtained removals at levels above 90% [21–25].

Concurrently with the reduction of ammonium nitrogen and nitrite nitrogen concentrations, a steady increase in nitrate nitrogen concentration is observed [4,12]. This form of nitrogen is safe for cultured fish, and concentrations as high as >70 mg N dm^{-3} have been tolerated by fish (see studies by Sikora et al. [18] and Steinberg et al. [26]). Although, Davidson et al. [27], for an *O. mykiss* culture in RAS, recommend nitrate nitrogen concentrations below 75 mg N dm^{-3}. However, a widely varying range of nitrate nitrogen concentrations is noted in RAS, with nitrate concentrations in closed systems (without partial water exchange) being higher and reaching levels above 50 mg N dm^{-3} [18,21,22] and even 150 mg N dm^{-3} [28]. On the other hand, in semi-closed systems (as tested in our study), concentrations are lower, at the level of several milligrams [18,20], because partial water exchange in these RAS is designed to remove nitrate nitrogen which accumulates in the system in the absence of denitrification [17,29].

3.2. Differences in N, P, and C Conversion between Bioreactors

The media plays an important function in filtration systems as the site of the physical processes (e.g., deposition of solid particles), as well as key biochemical processes leading to the transformation of substances present in the water, such as nitrification, during which, ammonium nitrogen and nitrite nitrogen are oxidized to the less toxic nitrate nitrogen [6,17,30]. In our study, as depicted by the linear regressions summarized in Figure 2, we noted consistent trends of decreasing concentrations of toxic forms of nitrogen and increasing concentrations of nitrate nitrogen. Though the conversion rate to nitrate nitrogen were similar in all bioreactors, those with BR-2 and BR-3 media were more effective in reducing nitrite and ammonium nitrogen concentrations. At the same time, the significantly ($p < 0.05$) lowest concentrations of both these forms of nitrogen were in the system with BR-3, followed by BR-2. In contrast, nitrate nitrogen concentrations did not significantly ($p > 0.05$) differ between the systems; although, on average, the lowest NO_3^--N concentration was recorded for BR-1 (Table 2). The reason for the differences may have been the specific surface area of the media. BR-1, with a smooth surface and the lowest SSA, was characterized by a lower degree of transformation of nitrogen compounds. However, it should be pointed out that the porous BR-2 medium (with the highest SSA) may have had its active surface area reduced to a greater extent by biofouling than BR-3, which, in consequence, may have lowered the efficiency of nitrogen transformation (although the differences in concentration of nitrogen forms between them were not significant).

The positive effect of an increase in SSA on nitrogen transformation has been shown, for example, by [7,31,32]. The larger surface area of a biofilter increases its functional efficiency, where an increase in the surface area colonized by bacteria increases the contact area of the biofilm with nutrients dissolved in the circulating water, and, thus, the efficiency of the biochemical processes of nitrogen conversion. The structure of the medium (e.g., the BR-3 medium is made of soft sponge) is also significant, and may promote a more intense flow of water through open and interconnected structures, leading to an increase in nitrification rates [5,8]. In contrast, the presence of excessively large voids with a simultaneously smooth surface (as may have been the case for the smooth BR-1 medium) may have resulted in a lower water renewal [33].

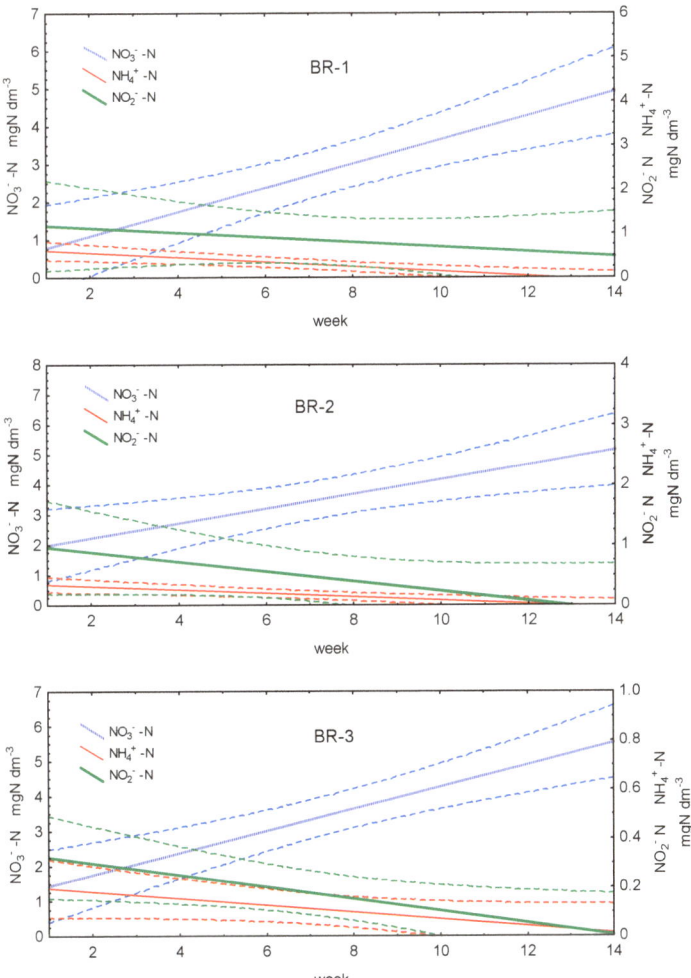

Figure 2. Linear regressions of the variation of inorganic forms of nitrogen in water of RAS systems with selected tessellated filter media, RK Plast (BR-1), Mutag-BioChip30 (BR-2), and LevaPor (BR-3).

It should also be emphasized that in the observed changes in the concentration of inorganic nitrogen forms, the retention on the biofilter media was low. These dissolved forms of nitrogen are retained only on nanofiltration membranes. For example, a 450 Da membrane in a RAS water treatment system reduced ammonia nitrogen by 44% and nitrate nitrogen by 9% [34].

For phosphorus and carbon, there was mostly no significant variation between the tested media. Only mean TRP and TP concentrations were significantly ($p < 0.05$) higher in the water filtered using BR-3 medium compared to the other systems. TIC concentrations did not differ between the tested media ($p > 0.05$), and mean TOC concentrations were significantly ($p < 0.05$) higher for BR-1 and BR-2 compared to BR-3 (Table 2). Overall, the demonstrated lack of reduction in P and C can be considered a positive phenomenon, as these compounds are not threatening factors for fish, and a deficit could negatively affect the desired biochemical processes of nitrogen transformation.

It should be emphasized that in the process of transformation of ammonium and nitrite nitrogen to nitrate nitrogen, a relatively low C/N ratio is desirable, as it provides the desired nitrate concentration in RAS [10,17,29]. In our study, TOC concentrations were at similar levels also recorded in other *O. mykiss* culture systems (e.g., Santorio et al. [35] recorded TOC concentrations of 8.14 mg C dm^{-3}), and the average TOC/TN values were 1.4 for BR-1, 1.3 for BR-2, and 1.1. for BR-3, and should be considered favorable. Indeed, Navada et al. [10], for example, showed a marked decrease in nitrification efficiency with an increase in the C/N ratio from 0 to 3. Similarly, Rojas-Tirado et al. [36] postulate that RAS effluent with a C/N ratio below 3, given a sufficient supply of oxygen, will ensure equilibrium in the bioreactor and rapid and stable nitrification.

4. Conclusions

Commercial media with different specific surface areas (700 m^2 m^{-3} for BR-1; 5500 m^2 m^{-3} for BR-2; 2700 m^2 m^{-3} for BR-3) were tested for passive activation capabilities in water treatment bioreactors in recirculating rainbow trout culture systems. For ammonium nitrogen and nitrite nitrogen, after increases in concentration during the first 2–3 weeks of culturing, a reduction in concentration to less than 0.1 mg N dm^{-3} was recorded. The reductions in ammonium nitrogen were 94.5% (BR-1), 91.8% (BR-2), and 92.3% (BR-3), and the reductions in nitrite nitrogen were 98.2% (BR-1 and BR-2) and 91.9% (BR-3). For nitrate nitrogen, there was a steady trend of increasing concentrations with the duration of the experiment to maximum concentrations (in mg N dm^{-3}) of 6.521 (BR-1), 6.778 (BR-2), and 7.326 (BR-3).

Efficient nitrification was recorded for each media variant, with the concentrations of nitrate nitrogen and the toxic forms of ammonium nitrogen and nitrite nitrogen at steady state levels safe for the cultured fish. The efficiency of nitrification depended on the specific surface area of the artificial media, which can be ranked in the following descending order of efficiency: BR-3 $\geq$ BR-2 > BR-1. The recorded concentrations of phosphorus and carbon allow us to conclude that their availability for microorganisms should not be a limiting factor in bioprocesses occurring in purification systems. At the same time, the low average TOC/TN ratio (1.4 for BR-1, 1.3 for BR-2, and 1.1 for BR-3) did not adversely affect nitrification.

The conducted research indicates the possible application of passive activation of the biofilter bed in recirculating fish farming systems with high environmental requirements, but further research is recommended to confirm this on other salmonid species. The presented results will be confronted in the future with the effects of biofilters working under conditions of technological culturing of *O. mykiss*.

Author Contributions: Conceptualization, A.N., M.B., A.T. (Agnieszka Tórz), A.T. (Adam Tański) and K.F.; Data curation, A.N., M.B., A.T. (Agnieszka Tórz) and A.T. (Adam Tański); Formal analysis, A.N. and A.T. (Agnieszka Tórz); Funding acquisition, K.F.; Investigation, A.N.; Methodology, A.N., M.B., A.T. (Adam Tański) and K.F.; Project administration, K.F.; Visualization, A.N. and A.T. (Adam Tański); Writing—original draft, A.N. All authors have read and agreed to the published version of the manuscript.

Funding: The study was conducted within the project no., 00002-6521.1-OR1600001/17/20, financed by Sectoral Operational Programme "Fisheries and See 2014-2020" and by the Polish Ministry of Science in Poland through a subsidy for the West Pomeranian University of Technology Szczecin, Faculty of Food Sciences and Fisheries.

Data Availability Statement: The data presented in this study are available on request from the corresponding author.

Conflicts of Interest: The authors declare no conflict of interest.

Appendix A

Table A1. Comparison of hydrochemical concentrations (mean values ± standard deviation) between water flowing into the bioreactor (inflow) and water flowing out of the bioreactor (outflow), and a summary of the results of one-way ANOVA (F and P).

Bioreactor		$NO_2^- $-N	$NO_3^- $-N	$NH_4^+ $-N	TN	TRP	TP	TIC	TOC
		mg N dm^{-3}				mg P dm^{-3}		mg C dm^{-3}	
BR-1	inflow	0.811 ±1.401	2.857 ±2.065	0.276 ±0.363	7.777 ±3.624	0.213 ±0.128	0.252 ±0.117	32.2 ±6.2	11.2 ±1.4
	outflow	0.845 ±1.42	2.850 ±2.072	0.266 ±0.347	7.803 ±3.554	0.214 ±0.130	0.251 ±0.119	32.5 ±6.0	11.0 ±1.2
	F	1.024	1.006	1.112	1.039	1.037	1.030	1.052	1.397
	P	0.949	0.999	0.936	0.984	0.990	0.991	0.902	0.720
BR-2	inflow	0.432 ±1.042	3.563 ±1.854	0.157 ±0.195	8.350 ±3.531	0.186 ±0.097	0.241 ±0.079	32.1 ±6.3	11.8 ±2.2
	outflow	0.445 ±1.157	3.578 ±1.990	0.159 ±0.224	8.315 ±3.512	0.190 ±0.099	0.250 ±0.078	31.7 ±6.3	11.4 ±1.7
	F	1.231	1.152	1.318	1.010	1.023	1.028	1.014	1.688
	P	0.976	0.983	0.987	0.994	0.931	0.941	0.887	0.569
BR-3	inflow	0.168 ±0.249	3.574 ±1.956	0.103 ±0.155	9.749 ±4.261	0.292 ±0.163	0.326 ±0.158	29.6 ±6.6	10.5 ±1.6
	outflow	0.162 ±0.252	3.418 ±1.928	0.107 ±0.185	9.673 ±4.260	0.294 ±0.164	0.327 ±0.158	29.6 ±6.6	10.1 ±1.7
	F	1.016	1.029	1.416	1.000	1.011	1.000	1.004	1.017
	P	0.946	0.834	0.949	0.963	0.964	0.995	0.965	0.606

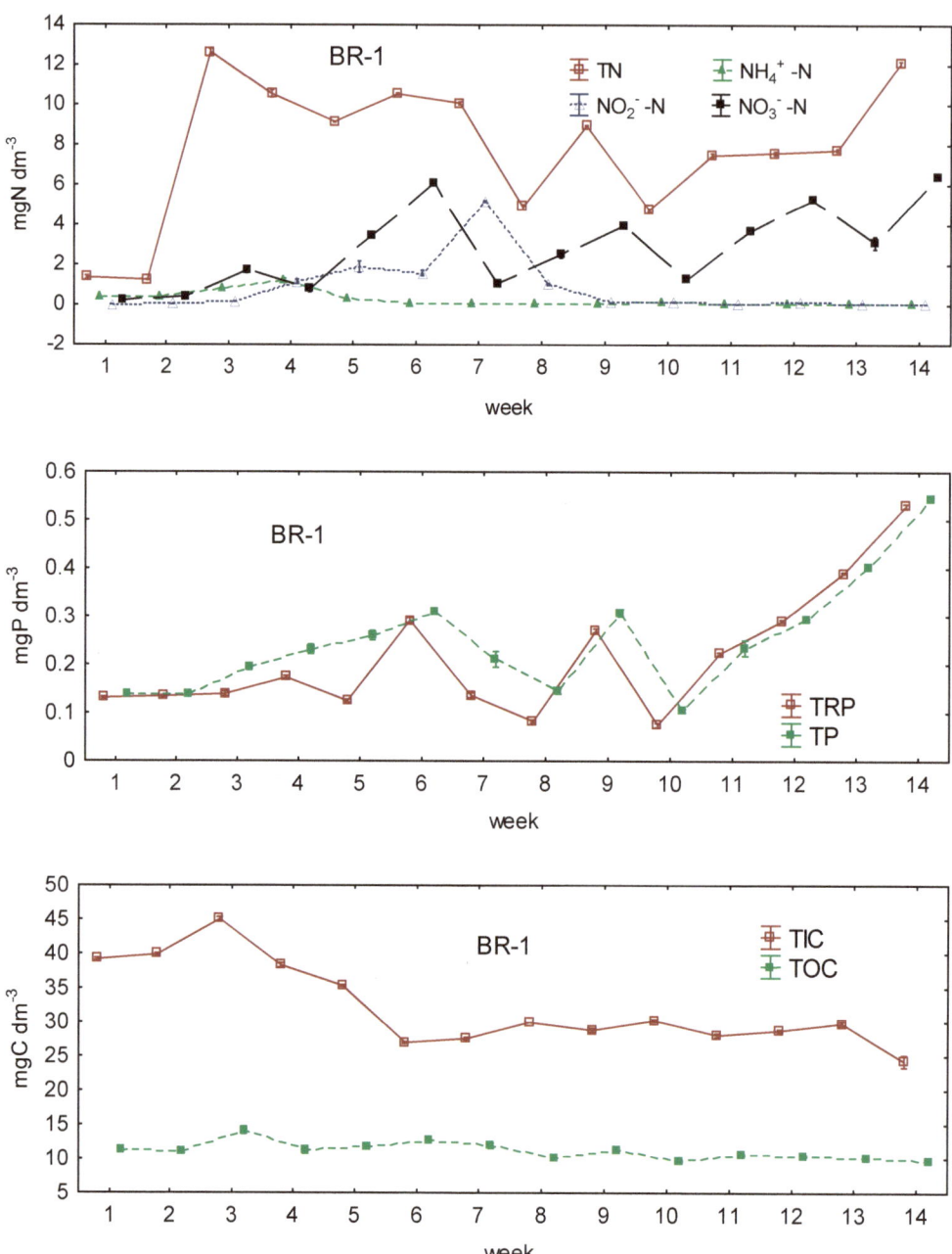

Figure A1. *Cont.*

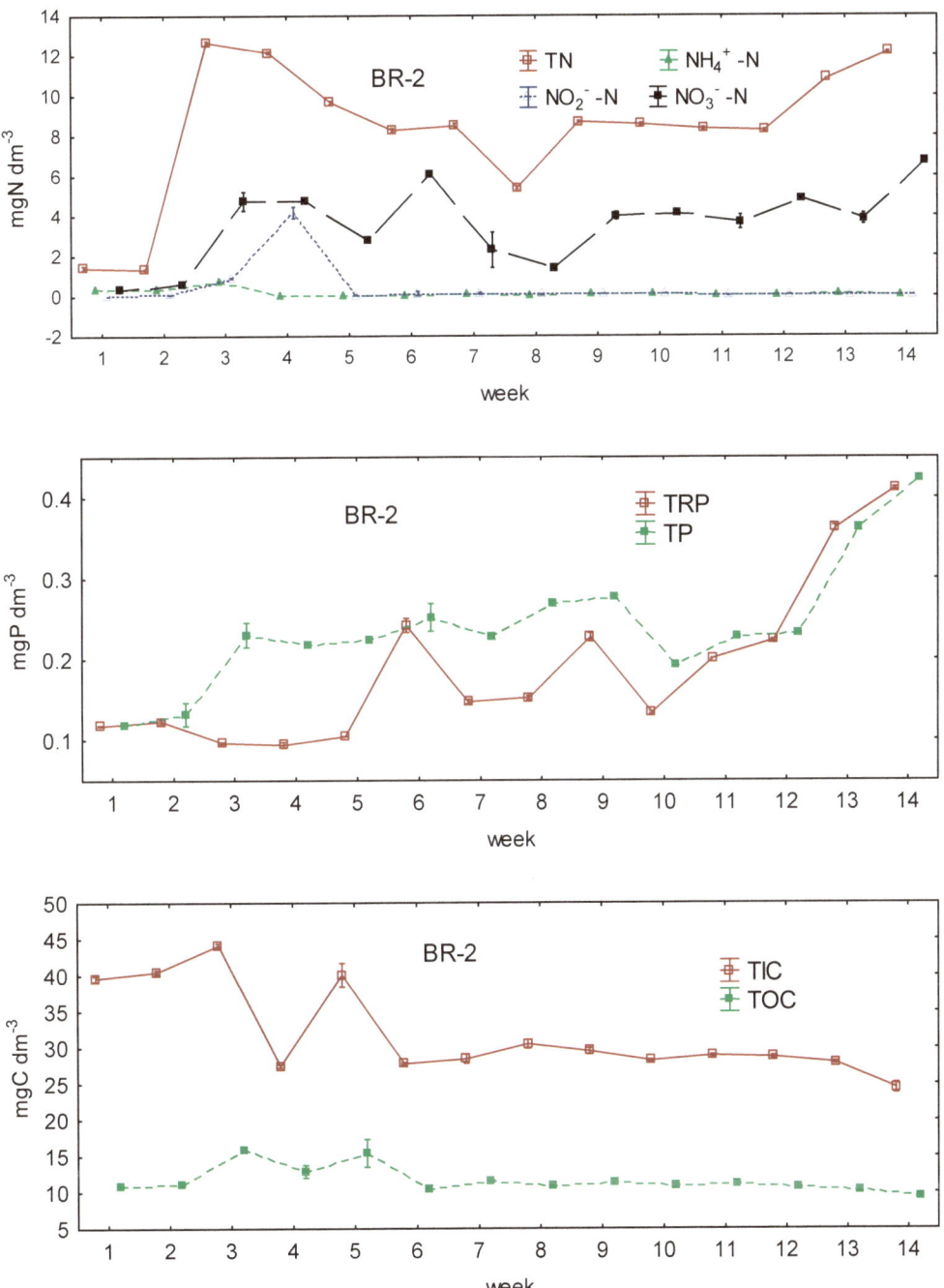

Figure A1. *Cont.*

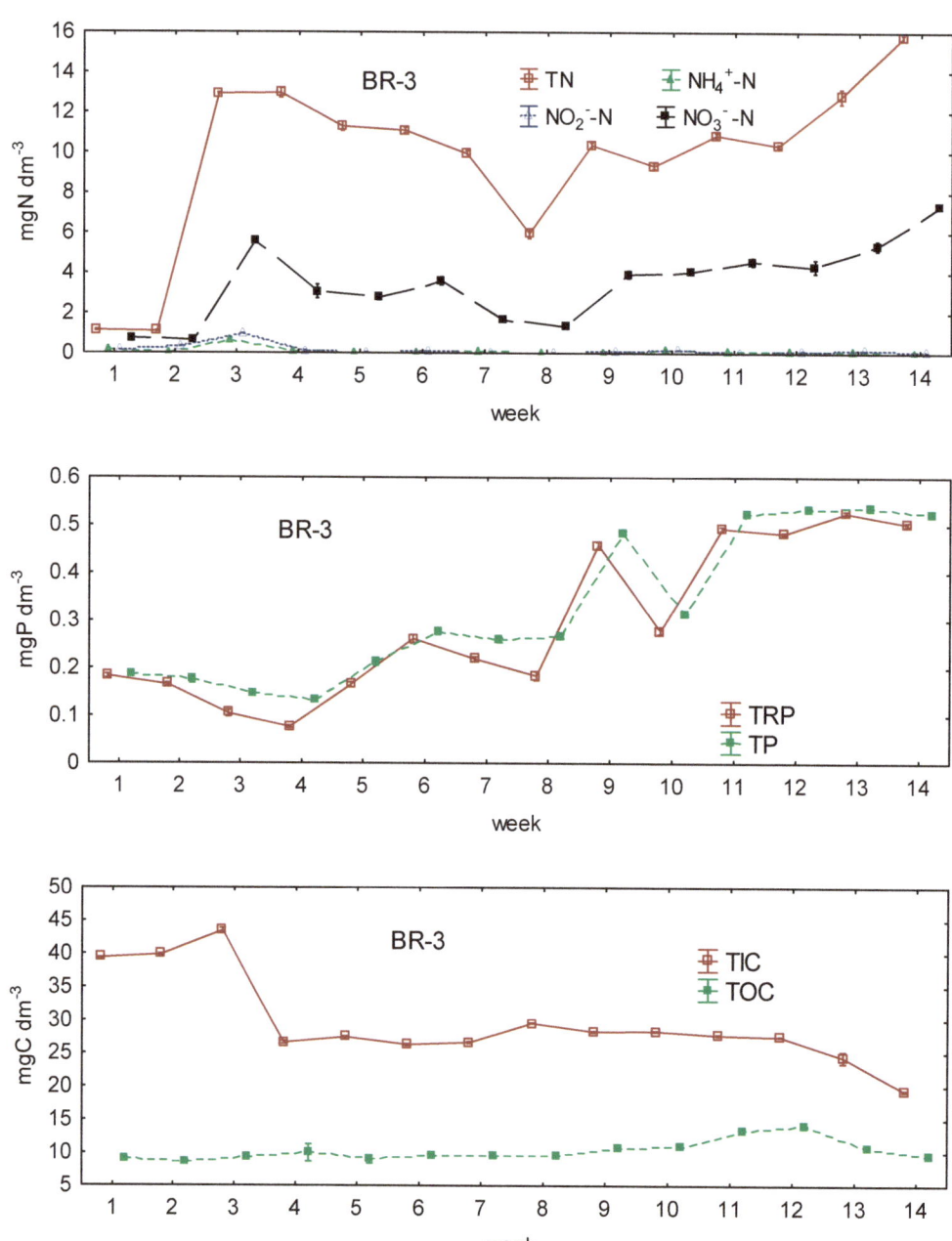

Figure A1. Changes in the concentration of the determinants of the water quality of RAS systems with selected tessellated RK Plast (BR-1), Mutag-BioChip30 (BR-2), and LevaPor (BR-3) filtration fillings (TN—nitrogen toxin; TRP—total reactive phosphorus; TP—total phosphorus; TIC—total inorganic carbon; TOC—total organic carbon) (square—mean; whisker—mean ± standard deviation).

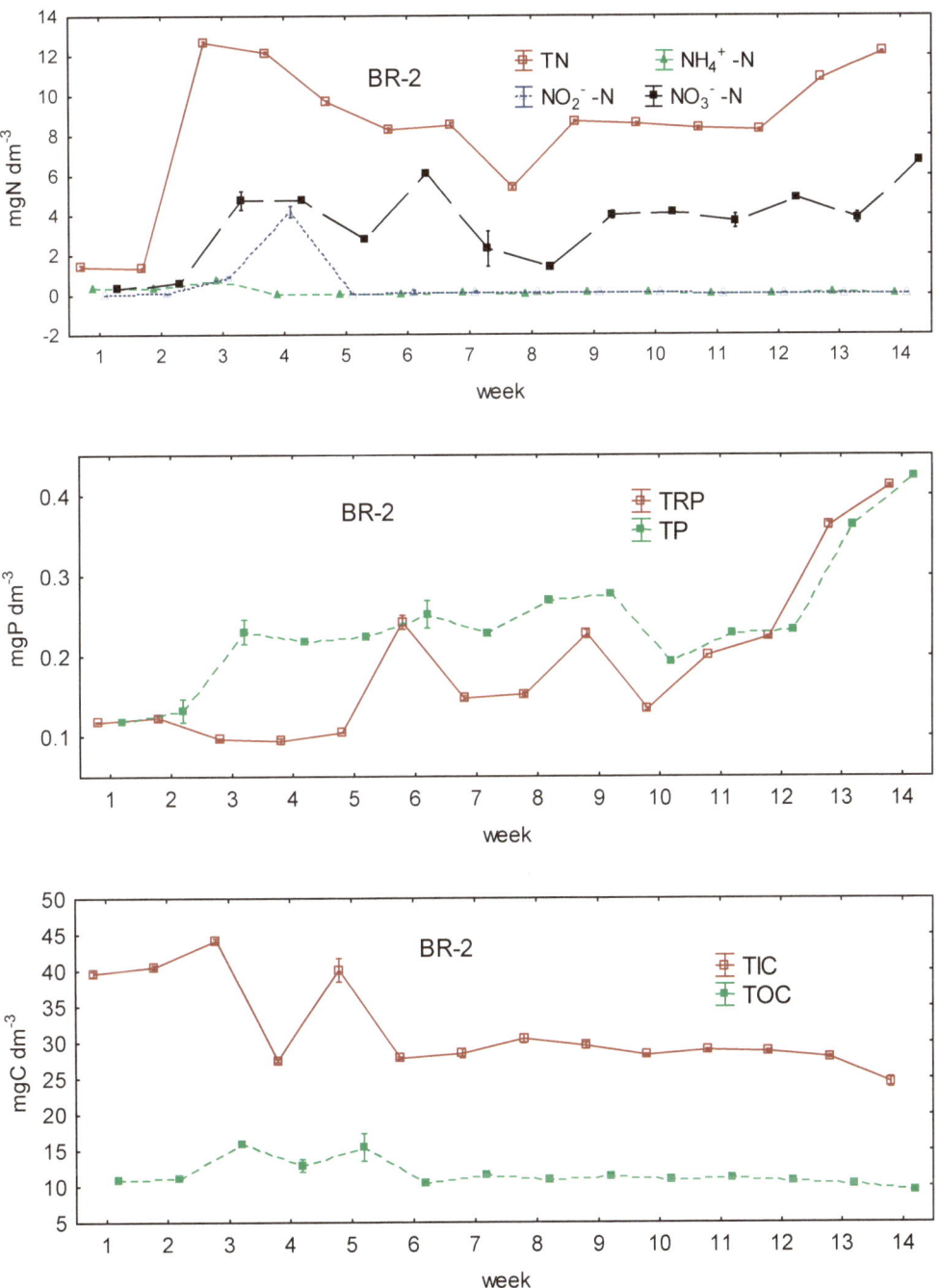

Figure A1. *Cont.*

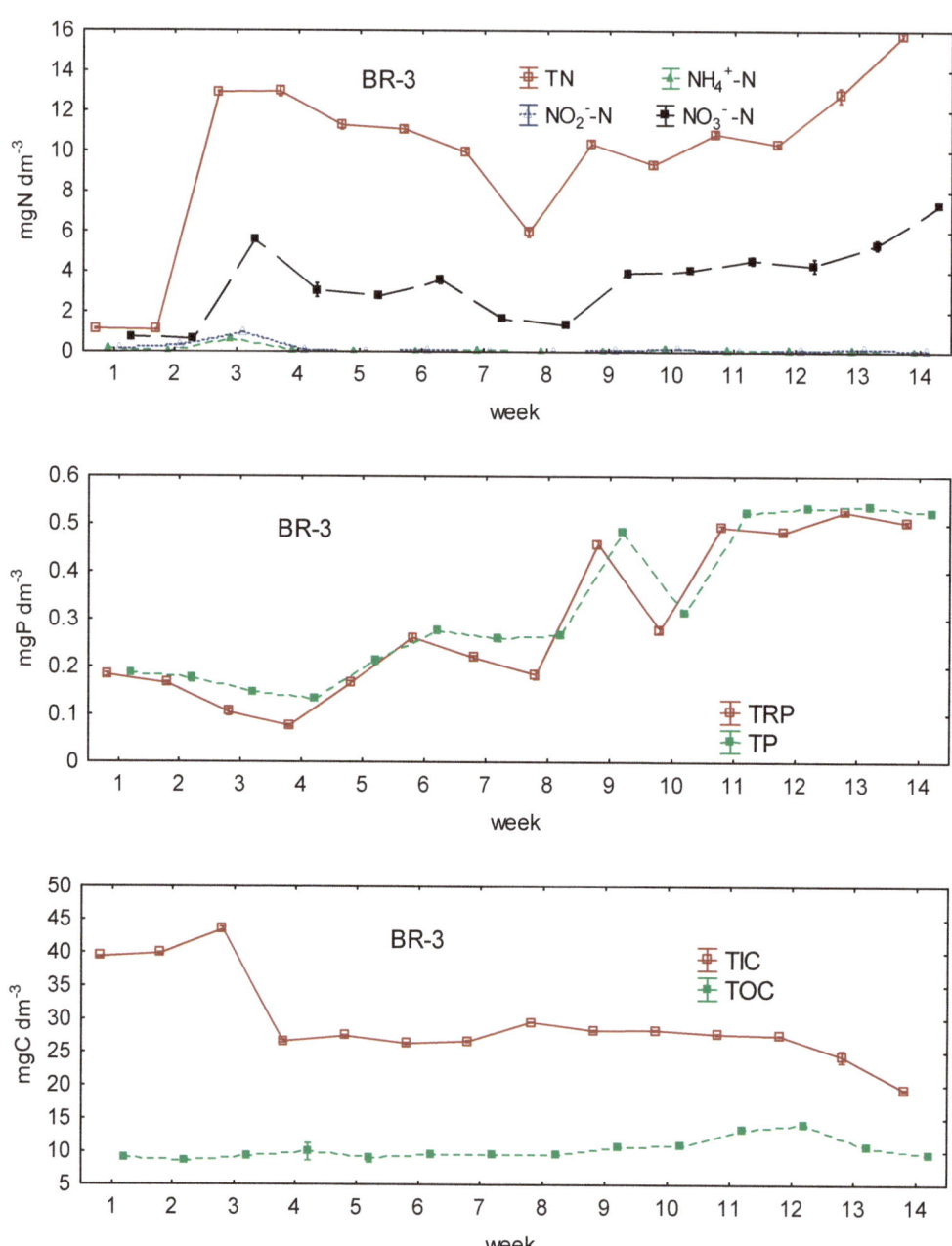

Figure A1. Changes in the concentration of the determinants of the water quality of RAS systems with selected tessellated RK Plast (BR-1), Mutag-BioChip30 (BR-2), and LevaPor (BR-3) filtration fillings (TN—nitrogen toxin; TRP—total reactive phosphorus; TP—total phosphorus; TIC—total inorganic carbon; TOC—total organic carbon) (square—mean; whisker—mean ± standard deviation).

References

1. FAO. *The State of World Fisheries and Aquaculture 2020: Sustainability in Action*; FAO: Rome, Italy, 2020. [CrossRef]
2. Sirakov, I. Flesh quality in rainbow trout (Oncorhynchus mykiss W.) and brown trout (Salmo trutta m. fario L.) cultivated in recirculation aquaculture system. *Int. J. Curr. Microbiol. Appl. Sci.* **2015**, *4*, 50–57.
3. Wang, C.; Li, Z.; Wang, T.; Xu, X.; Zhang, X.; Li, D. Intelligent fish farm—The future of aquaculture. *Aquac. Int.* **2021**, *29*, 2681–2711. [CrossRef]
4. Minaz, M.; Kubilay, A. Operating parameters affecting biofloc technology: Carbon source, carbon/nitrogen ratio, feeding regime, stocking density, salinity, aeration, and microbial community manipulation. *Aquac. Int.* **2021**, *29*, 1121–1140. [CrossRef]
5. Zou, Y.; Hu, Z.; Zhang, J.; Xie, H.; Guimbaud, C.; Fang, Y. Effects of pH on nitrogen transformations in media-based aquaponics. *Bioresour. Technol.* **2016**, *210*, 81–87. [CrossRef]
6. Zhu, S.-M.; Deng, Y.-L.; Ruan, Y.-J.; Guo, X.S.; Shi, M.M.; Shen, J.Z. Biological denitrification using poly(butylene succinate) as carbon source and biofilm carrier for recirculating aquaculture system effluent treatment. *Bioresour. Technol.* **2015**, *192*, 603–610. [CrossRef]
7. Ridha, M.T.; Cruz, E.M. Effect of biofilter media on water quality and biological performance of the Nile tilapia *Oreochromis niloticus* L. reared in a simple recirculating system. *Aquac. Eng.* **2001**, *24*, 157–166. [CrossRef]
8. Boaventura, T.P.; Miranda-Filho, K.C.; Oréfice, R.L.; Luz, R.K. Influence of porosity of low-density polyethylene media on the maturation process of biofilters used in recirculating aquaculture systems. *Aquac. Int.* **2018**, *26*, 1035–1049. [CrossRef]
9. DeLong, D.P.; Losardo, T.M. *How to Start a Biofilter*; United States Department of Agriculture Southern Regional Aquaculture Center: Stoneville, MS, USA, 2012; p. 4.
10. Navada, S.; Vadstein, O.; Spanu, C.; Mikkelsen, Ø.; Kolarevic, J. Biofilms remember: Osmotic stress priming as a microbial management strategy for improving salinity acclimation in nitrifying biofilms. *Water Res.* **2020**, *176*, 115732. [CrossRef]
11. Li, C.; Liang, J.; Lin, X.; Xu, H.; Tadda, M.A.; Lan, L. Fast start-up strategies of MBBR for mariculture wastewater treatment. *J. Environ. Manage.* **2019**, *248*, 109267. [CrossRef]
12. APHA. *Standard Methods for Examination of Water and Wastewater*, 20th ed.; American Public Health Association: Washington, DC, USA, 1999; p. 1325. ISBN 978-0875532356.
13. Dias, J.; Bellingham, M.; Hassan, J.; Barrett, M.; Stephenson, T.; Soares, A. Influence of carrier media physical properties on start-up of moving attached growth systems. *Bioresour. Technol.* **2018**, *266*, 463–471. [CrossRef]
14. Masser, M.P.; Rakocy, J.; Losardo, T.M. *Recirculating Aquaculture Tank Production Systems*; United States Department of Agriculture Southern Regional Aquaculture Center: Stoneville, MS, USA, 1999; p. 10.
15. Avnimelech, Y.; Ritvo, G. Shrimp and fish pond soils: Processes and management. *Aquaculture* **2003**, *220*, 549–567. [CrossRef]
16. Schneider, O.; Sereti, V.; Eding, E.H.; Verreth, J.A.J. Molasses as C source for heterotrophic bacteria production on solid fish waste. *Aquaculture* **2006**, *261*, 1239–1248. [CrossRef]
17. Ruiz, P.; Vidal, J.M.; Sepulveda, D.; Torres, C.; Villouta, G.; Carrasco, C.; Aguilera, F.; Ruiz-Tagle, N.; Urrutia, H. Overview and future perspectives of nitrifying bacteria on biofilters for recirculating aquaculture systems. *Rev. Aquac.* **2020**, *12*, 1478–1494. [CrossRef]
18. Sikora, M.; Nowosad, J.; Biegaj, M.; Kucharczyk, D.; Dębowski, M. The possibility of application of agglomerate elastomers (EPP) as media for biological bed in aquaculture. *Aquac. Res.* **2018**, *49*, 2988–2994. [CrossRef]
19. Owatari, M.S.; Jesus, G.F.A.; de Melo Filho, M.E.S.; Lapa, K.R.; Martins, M.L.; Mouriño, J.L.P. Synthetic fibre as biological support in freshwater recirculating aquaculture systems (RAS). *Aquac. Eng.* **2018**, *82*, 56–62. [CrossRef]
20. Żarski, D.; Kucharczyk, D.; Targońska, K.; Chyła, B.; Dobrołowicz, A. Dynamics of changes in nitrogen and phosphorus compounds during intensive rearing of ide, Leuciscus idus (L.), in a recirculating system. *Arch. Polish Fish.* **2008**, *16*, 459–467. [CrossRef]
21. Pulkkinen, J.T.; Eriksson-Kallio, A.M.; Aalto, S.L.; Tiirola, M.; Koskela, J.; Kiuru, T.; Vielma, J. The effects of different combinations of fixed and moving bed bioreactors on rainbow trout (Oncorhynchus mykiss) growth and health, water quality and nitrification in recirculating aquaculture systems. *Aquac. Eng.* **2019**, *85*, 98–105. [CrossRef]
22. Fernandes, P.M.; Pedersen, L.; Pedersen, P.B. Influence of fixed and moving bed biofilters on micro particle dynamics in a recirculating aquaculture system. *Aquac. Eng.* **2017**, *78*, 32–41. [CrossRef]
23. Pulkkinen, J.T.; Kiuru, T.; Aalto, S.L.; Koskela, J.; Vielma, J. Startup and effects of relative water renewal rate on water quality and growth of rainbow trout (Oncorhynchus mykiss) in a unique RAS research platform. *Aquac. Eng.* **2018**, *82*, 38–45. [CrossRef]
24. Shitu, A.; Zhu, S.; Qi, W.; Abubakar, M.A.; Liu, D. Performance of novel sponge biocarrier in MBBR treating recirculating aquaculture systems wastewater: Microbial community and kinetic study. *J. Environ. Manag.* **2020**, *275*, 111264. [CrossRef]
25. Fu, H.; Wang, J.; Ren, H.; Ding, L. Acceleration of start-up of moving bed biofilm reactor at low temperature by adding specialized quorum sensing bacteria. *Bioresour. Technol.* **2022**, *358*, 127249. [CrossRef] [PubMed]
26. Steinberg, K.; Zimmermann, J.; Meyer, S.; Schulz, C. Start-up of recirculating aquaculture systems: How do water exchange rates influence pikeperch (Sander lucioperca) and water composition? *Aquac. Eng.* **2018**, *83*, 151–159. [CrossRef]
27. Davidson, J.; Good, C.; Welsh, C.; Summerfelt, S.T. Comparing the effects of high vs. low nitrate on the health, performance, and welfare of juvenile rainbow trout Oncorhynchus mykiss within water recirculating aquaculture systems. *Aquac. Eng.* **2014**, *59*, 30–40. [CrossRef]

28. Stavrakidis-Zachou, O.; Ernst, A.; Steinbach, C.; Wagner, K.; Waller, U. Development of denitrification in semi-automated moving bed biofilm reactors operated in a marine recirculating aquaculture system. *Aquac. Int.* **2019**, *27*, 1485–1501. [CrossRef]
29. He, Q.; Zhang, D.; Main, K.; Feng, C.; Ergas, S.J. Biological denitrification in marine aquaculture systems: A multiple electron donor microcosm study. *Bioresour. Technol.* **2018**, *263*, 340–349. [CrossRef] [PubMed]
30. Li, W.; Shi, C.; Yu, Y.; Ruan, Y.; Kong, D.; Lv, X.; Xu, P.; Awasthi, M.K.; Dong, M. Interrelationships between tetracyclines and nitrogen cycling processes mediated by microorganisms: A review. *Bioresour. Technol.* **2021**, *319*, 124036. [CrossRef] [PubMed]
31. Nguyen, T.T.; Ngo, H.H.; Guo, W.; Johnston, A.; Listowski, A. Effects of sponge size and type on the performance of an up-flow sponge bioreactor in primary treated sewage effluent treatment. *Bioresour. Technol.* **2010**, *101*, 1416–1420. [CrossRef]
32. Massoompour, A.R.; Borghei, S.M.; Raie, M. Enhancement of biological nitrogen removal performance using novel carriers based on the recycling of waste materials. *Water Res.* **2020**, *170*, 115340. [CrossRef]
33. Chen, X.; Kong, L.; Wang, X.; Tian, S.; Xiong, Y. Accelerated start-up of moving bed biofilm reactor by using a novel suspended carrier with porous surface. *Bioprocess Biosyst. Eng.* **2015**, *38*, 273–285. [CrossRef]
34. Bonisławska, M.; Nędzarek, A.; Rybczyk, A. Assessment of the use of precipitating agents and ceramic membranes for treatment of effluents with high concentrations of nitrogen and phosphorus from recirculating aquaculture systems. *Aquac. Res.* **2019**, *50*, 1248–1256. [CrossRef]
35. Santorio, S.; del Rio, A.V.; Amorim, C.L.; Arregui, L.; Castro, P.M.L.; Mosquera-Corral, A. Pilot-scale continuous flow granular reactor for the treatment of extremely low-strength recirculating aquaculture system wastewater. *J. Environ. Chem. Eng.* **2022**, *10*, 107247. [CrossRef]
36. Rojas-Tirado, P.; Bovbjerg, P.; Vadstein, O.; Pedersen, L. Microbial dynamics in RAS water: Effects of adding acetate as a biodegradable carbon-source. *Aquac. Eng.* **2019**, *84*, 106–116. [CrossRef]

Article

Characterization of Dioxins and Heavy Metals in Chelated Fly Ash

Shuping Pan [1,2], Qi Yao [3], Wenxiang Cai [1,2,*], Yaqi Peng [3], Yuhao Luo [1], Zhizhen Wang [1], Caiping Jiang [1], Xiaodong Li [3] and Shengyong Lu [3]

[1] Zhejiang Ecological Environment Monitoring Center, Hangzhou 310007, China; panshuping@zjemc.org.cn (S.P.); luoyuhao@zjemc.org.cn (Y.L.); wangzhizhen@zjemc.org.cn (Z.W.); jiangcaiping@zjemc.org.cn (C.J.)

[2] Zhejiang Key Laboratory of Ecological Environment Monitoring Forewarning and Quality Control, Hangzhou 310007, China

[3] Institute for Thermal Power Engineering, Zhejiang University, Hangzhou 310027, China; yaoq@zju.edu.cn (Q.Y.); pengyaqi@zju.edu.cn (Y.P.); lixd@zju.edu.cn (X.L.); lushy@zju.edu.cn (S.L.)

* Correspondence: caiwenxiang@zjemc.org.cn

Abstract: Municipal solid waste incineration (MSWI) fly ash contains highly toxic heavy metals and polychlorinated dibenzo dioxins/furans (PCDD/Fs), which are a type of hazardous waste. The pollution characteristics of fly ash have changed with the development of stoker grate incinerators and the fly ash treatment technology; however, no research has been focused on this in recent years. In this study, 12 fly ash samples were collected from 9 grate power plants in southeastern China, and their PCDD/Fs and heavy metal concentrations were determined and compared to previous fly ash data. The PCDD/Fs concentration in fly ash was in the range of 0.002–0.051 ngI-TEQ/g, with an average of 0.027 ngI-TEQ/g. Furthermore, 1,2,3,4,6,7,8-HpCDD and OCDD made the most significant contributions to PCDDs. The distribution of 10 dioxins exhibited bimodal, unimodal, and normal characteristics. Linear fitting demonstrated a strong correlation between toxicity and 1,2,3,7,8-PentaCDD, 1,2,3,7,8-PentaCDF, and 2,3,4,7,8-PentaCDF. Concerning heavy metals, Pb poses a significant environmental risk. This is the first time that fly ash treated with a chelating agent has been thoroughly analyzed, which is vital for understanding the pollution level and treatment of fly ash derived from current power plants.

Keywords: MSWI fly ash; PCDD/Fs; heavy metals

1. Introduction

China's population has increased rapidly since the turn of the century, as has the level of urbanization, which has been accompanied by a rapid increase in waste production [1]. In 2020, China removed 235 million tons of waste. Landfilling is an unsustainable method of waste disposal that has expanded into the limited space along China's eastern coast [2]. China has made significant efforts to build waste incineration plants in order to address the phenomenon of "garbage siege", and its treatment capacity is increasing yearly. There are currently 463 harmless waste incineration plants in China, and waste incineration has surpassed landfills as the main from of waste disposal [3]. Municipal solid waste incineration (MSWI) effectively reduces waste, is harmless, and is internationally recognized as an advanced method of waste disposal [4]. However, MSWI fly ash has become a new pollutant. Because the dioxins and heavy metals in fly ash pose a serious threat to human health, a life cycle assessment is required to determine the long-term pollution characteristics of fly ash in the surrounding area [5].

Polychlorinated dibenzo dioxins/furans (PCDD/Fs) are highly toxic, and one of the most significant sources of these compounds is waste combustion [6]. Countries have long implemented policies that limit dioxin emissions and optimize combustion processes to reduce dioxin generation at the source [7]. As a result of the highly polluting properties of fly ash, current related research hotspots include heat treatment, hydrothermal treatment, curing/stabilization [8], the microwave method, and photocatalysis [9–11]. In practice, treatment with fly ash chelating agents is a low-cost method. Currently, fly ash chelating agents include diacetic acid, phosphate, and dithiamine, among others, which eventually form water-insoluble, stable polymeric heavy metal ion chelates. At present, this is the primary method for treating fly ash containing heavy metals [12]. The concentration of dioxins in fly ash disposed of in landfills must meet the standard of 3 ng/g, as specified in GB16889-2008; however, for grate furnaces, the content of dioxins in fly ash is not high, and this threshold is easy to meet with the development of municipal solid waste flue gas treatment technology.

The distribution of dioxin content in fly ash is determined by waste composition, waste incineration technology, and fly ash treatment technology, among other factors. The different compositions of dioxins are very important in terms of understanding the synthetic process and the toxicity level of dioxins. Dioxins are formed in three ways: high-temperature synthesis, ab initio synthesis, and precursor synthesis [13]. Fly ash is the primary reaction surface for the formation of dioxins. The temperature distribution of flue gas in MSWI is the main reason for the formation of dioxin. The rapid cooling of flue gas below 260 °C can reduce the production of dioxins. Using a separator to separate fly ash and flue gas within a high-temperature area is also an effective means to reduce the formation of dioxins [7]. The distribution of dioxins in fly ash can be used to infer the dioxin formation process [14], which aids in the development of relevant emission standards [15].

For many years, scholars have been investigating the characteristic distribution of dioxins in China's MSWI fly ash [16,17]; however, the scope is broad, a long time has passed, and the overall pollution level of fly ash in a specific area cannot be accurately represented. In this study, 14 sets of emission data from 9 grate MSWI power plants in southeastern China were analyzed and compared with previous fly ash data from southeastern China. An air pollution control device for selective non-catalytic reduction (SNCR) denitrification + semi-dry deacidification + activated carbon adsorption + cloth-bag dedusting was adopted, and the fly ash was chelated with the polymer dithioamine. Using these data, the distribution of dioxins in grate furnaces in China was investigated. Moreover, the distribution of homologues, their main contribution and correlation to dioxin toxicity, and the heavy metal data of various fly ash samples were assessed.

2. Materials and Methods

2.1. Basic Situation of Sampling Power Plants

The fly ash was collected from 12 groups of grate furnace data from 9 MSWI power plants in a specific area of southeast China. Moreover, previous grate furnace research data from southeast China were collected [16]. The fly ash samples were labeled (GFA1, GFA2, and GFA12 denote data from this study, and the serial numbers PFA1, PFA2, and PFA6 denote data from previous research), and plant numbers are indicated by Arabic numerals (1, 2, and 15). The capacity (the tonnage of waste disposed of per day) is shown in Table 1. In this study, 12 groups of fly ash were collected from various factory ash hoppers, treated with chelating agents (dithioamine polymer and agents primarily chelating heavy metals), and tested for dioxin. Following testing, qualified fly ash was sent to a landfill for landfill treatment.

Table 1. Power plants and sampling.

Sample Number	Power Plant	Capacity	Sample Number	Power Plant	Capacity
GFA1	1	3 × 350 ton/d	GFA10	7	3 × 225 ton/d
GFA2			GFA11	8	2 × 225 ton/d
GFA3	2	2 × 500 ton/d	GFA12	9	3 × 350 ton/d
GFA4			PFA1 *	10	1000 ton/d
GFA5	3	2 × 600 ton/d	PFA2	11	2 × 600 ton/d
GFA6			PFA3	12	3 × 350 ton/d
GFA7	4	400 ton/d	PFA4	13	1500 ton/d
GFA8	5	4 × 750 ton/d	PFA5	14	1000 ton/d
GFA9	6	2 × 400 ton/d	PFA6	15	600 ton/d

* data from previous literature [16].

2.2. Sample Extraction and Analysis

The dioxin determination method for the samples was HJ 77.3-2008 isotope dilution high-resolution gas chromatography–high resolution mass spectrometry. The fly ash samples were treated with hydrochloric acid, and then washed and dried with filtered water before being extracted by dichloromethane oscillation and combined with toluene to form the extracts. The extracts were purified using a multi-layer silica gel column and separated using an activated carbon silica column for instrument analysis. The instrument model was Water AutoSpec Premier(Waters, Manchester, UK), which uses an electron-impact (EI) ion source and selective ion monitoring (SIM). The temperature program for chromatographic separation was as follows: from 130 °C (1 min) to 210 °C at 15 °C/min, then at 3 °C/min to 310 °C; then, the temperature was maintained for 8 min at 310 °C. For more details, please refer to previous research [18] and China Eco-environmental Standards HJ 77.3-2008 (Solid Waste Determination of PCDDs and PCDFs Isotope Dilution HRGC–HRMS).

Heavy metal analysis was performed using the Chinese national standard HJ/T300-2007 acetic acid buffer solution method. Inductively coupled plasma emission spectrometry HJ781-2016 (Optima 8300, PE, Singapore City, Singapore) were used for Cd, Ba, Cr, Ni, Pb, Cu, Be, and Zn analysis. Microwave digestion/atomic fluorescence HJ702-2014 (AFS-9103, Titan Instruments, Beijing, China) were used for Se and As analysis. Cold atomic absorption spectrophotometry GB/T15555.1-1995 (RA-915M, LUMEX, Vancouver, BC, Canada) were utilized for Hg, and Cr^{6+} was determined using the biphenylcarbonyl dihydrazine spectrophotometric GB/T15555.4-1995 method (TU-1810, Purkinje General, Beijing, China).

3. Results

3.1. Dioxin Emission Concentration and TEQ Level in Fly Ash

Table 2 shows the concentrations of 17 toxic dioxins and I-TEQ, 10 dioxin and furan homologues, the ratio of PCDD to PCDF, and the total amount of dioxins in the 18 samples. As can be seen in the table, the I-TEQ levels of the 12 grate fly ash samples in this study were in the range of 0.002–0.051 ngI-TEQ/g and 0.027 ngI-TEQ/g on average. The equivalent toxicity in Taiwan was reported to be 0.78–2.86 ngI-TEQ/g [19]. Moreover, the equivalent toxicity previously reported in China was 0.034–2.5 ngI-TEQ/g. Six types of grate fly ash from southeast China exhibited an I-TEQ concentration range from 0.115 to 0.645 ngI-TEQ/g [16]. The ranges in this study are much lower than those previously reported, which is largely due to the rapid development of grate furnaces and dioxin control technology in China. According to the Chinese landfill standard, all the samples met the limit of 3 ng/g, as specified in GB16889-2008.

Table 2. Data for PCDD/Fs for 18 samples (in ng/g).

PCDD/Fs	GFA1	GFA2	GFA3	GFA4	GFA5	GFA6	GFA7	GFA8	GFA9	GFA10	GFA11	GFA12	PFA1	PFA2	PFA3	PFA4	PFA5	PFA6
2,3,7,8-TetraCDD	0.004	0.008	0.004	0.003	0.001	0.001	0.002	0.003	0.001	0.000	0.003	0.001	0.039	0.0087	0.037	0.044	0.046	0.04
1,2,3,7,8-PentaCDD	0.010	0.018	0.014	0.010	0.003	0.005	0.005	0.005	0.002	0.000	0.010	0.004	0.11	0.02	0.14	0.15	0.11	0.11
1,2,3,4,7,8-HexaCDD	0.011	0.022	0.017	0.014	0.004	0.006	0.006	0.003	0.003	0.001	0.011	0.005	0.086	0.021	0.13	0.14	0.066	0.067
1,2,3,6,7,8-HexaCDD	0.026	0.057	0.028	0.021	0.008	0.013	0.019	0.020	0.009	0.002	0.024	0.013	0.64	0.075	0.27	0.42	0.13	0.13
1,2,3,7,8,9-HexaCDD	0.017	0.038	0.022	0.018	0.005	0.008	0.012	0.010	0.006	0.001	0.014	0.008	0.36	0.044	0.18	0.24	0.1	0.1
1,2,3,4,6,7,8-HeptaCDD	0.193	0.459	0.172	0.130	0.056	0.072	0.219	0.084	0.100	0.012	0.142	0.115	5	0.7	1.9	2.9	1.1	0.91
OCDD	0.291	0.843	0.248	0.167	0.103	0.094	0.632	0.267	0.403	0.041	0.252	0.408	7.2	1.4	4.8	5.2	2.4	1.7
2,3,7,8-TetraCDF	0.027	0.043	0.021	0.015	0.006	0.009	0.004	0.023	0.002	0.001	0.018	0.006	0.28	0.062	0.17	0.3	0.3	0.17
1,2,3,7,8-PentaCDF	0.027	0.061	0.027	0.019	0.011	0.018	0.008	0.019	0.005	0.001	0.031	0.009	0.42	0.099	0.44	0.55	0.48	0.41
2,3,4,7,8-PentaCDF	0.017	0.076	0.033	0.023	0.010	0.015	0.007	0.019	0.010	0.001	0.022	0.007	0.3	0.072	0.37	0.43	0.32	0.27
1,2,3,4,7,8-HexaCDF	0.029	0.067	0.031	0.020	0.020	0.031	0.019	0.011	0.012	0.002	0.036	0.016	0.35	0.076	0.47	0.42	0.35	0.42
1,2,3,6,7,8-HexaCDF	0.031	0.083	0.035	0.023	0.022	0.034	0.024	0.015	0.011	0.002	0.042	0.018	0.35	0.075	0.44	0.43	0.33	0.42
2,3,4,6,7,8-HexaCDF	0.041	0.115	0.028	0.021	0.032	0.048	0.048	0.017	0.018	0.003	0.046	0.023	0.25	0.079	0.39	0.37	0.21	0.31
1,2,3,7,8,9-HexaCDF	0.006	0.057	0.012	0.008	0.005	0.007	0.007	0.002	0.006	0.000	0.007	0.004	0.055	0.0096	0.04	0.051	0.038	0.04
1,2,3,4,6,7,8-HeptaCDF	0.077	0.234	0.092	0.057	0.073	0.095	0.120	0.016	0.059	0.006	0.097	0.057	0.61	0.23	1.3	1	0.6	1.1
1,2,3,4,7,8,9-HeptaCDF	0.011	0.052	0.017	0.010	0.013	0.017	0.034	0.003	0.008	0.001	0.016	0.013	0.15	0.05	0.17	0.16	0.1	0.24
OCDF	0.027	0.120	0.043	0.025	0.037	0.035	0.194	0.005	0.033	0.003	0.032	0.039	0.26	0.19	0.48	0.35	0.2	1.1
Total-tetradioxins	0.090	0.384	0.735	0.725	0.037	0.030	0.031	0.180	0.033	0.006	0.341	0.056	2.4	0.33	1.7	3.2	0.9	0.66
Total-pentadioxins	0.180	0.673	0.716	0.580	0.069	0.079	0.072	0.265	0.053	0.011	0.282	0.108	5.2	0.66	2.9	4.7	1.5	1.2
Total-hexadioxins	0.605	1.236	0.729	0.627	0.123	0.182	0.263	0.410	0.133	0.023	0.484	0.211	14	1.7	6	6.9	3	1.8
Total-heptadioxins	0.397	0.936	0.338	0.252	0.125	0.162	0.442	0.191	0.216	0.024	0.346	0.236	11	1.4	4.3	6	2.2	1.9
Total-OCDD	0.291	0.843	0.248	0.167	0.103	0.094	0.632	0.267	0.403	0.041	0.252	0.408	7.2	1.4	4.8	5.2	2.4	1.7
Total-PCDDs	1.563	4.073	2.767	2.351	0.458	0.548	1.441	1.312	0.838	0.104	1.704	1.019	39	5.6	20	26	10	7.3
Total-tetrafurans	0.188	0.695	0.413	0.264	0.111	0.111	0.058	0.413	0.051	0.009	0.374	0.090	7.1	1.7	6.2	11	7.3	5.2
Total-pentafurans	0.185	0.933	0.446	0.294	0.227	0.285	0.147	0.310	0.103	0.018	0.489	0.171	4.8	1.2	5.6	7	5.3	4.8
Total-hexafurans	0.171	0.473	0.160	0.110	0.220	0.334	0.231	0.080	0.071	0.018	0.234	0.146	2.9	0.71	4.1	3.8	2.8	3.6
Total-heptafurans	0.118	0.391	0.141	0.089	0.119	0.156	0.222	0.029	0.085	0.010	0.152	0.097	1.2	0.42	2	1.6	1	2.1
Total-OCDF	0.027	0.120	0.043	0.025	0.037	0.035	0.194	0.005	0.033	0.003	0.032	0.038	0.26	0.19	0.48	0.35	0.2	1.1
Total-PCDFs	0.688	2.612	1.204	0.782	0.714	0.921	0.854	0.837	0.342	0.058	1.281	0.542	16	4.2	18	24	17	17
Total(PCDDs+PCDFs)	2.251	6.685	3.970	3.134	1.172	1.469	2.295	2.149	1.180	0.162	2.985	1.561	55	9.8	38	50	27	24
PCDDs/PCDFs ratio	2.272	1.559	2.299	3.005	0.641	0.594	1.688	1.567	2.450	1.782	1.331	1.880	2.438	1.333	1.111	1.083	0.589	0.4294
I-TEQ	0.041	0.011	0.051	0.036	0.020	0.029	0.026	0.027	0.016	0.002	0.043	0.018	0.567	0.115	0.562	0.645	0.458	0.442

3.2. Toxic Dioxin Distribution in the Samples

The concentration and percentage concentration distribution of 17 toxic dioxins in 18 sample groups are shown in Figure 1. The difference in concentration in this study as compared to a previous study in vast. It can be seen in Figure 1a,c that the distribution of toxic dioxins in the grate furnace samples in this study was at a low level, especially much lower than the value shown in Figure 1e. In Figure 1b–f, it can also be seen that the main contributors to PCDD/Fs were consistent, but their relative content was different. Furthermore, 1,2,3,4,6,7,8-HpCDD and OCDD contributed the most to PCDDs, with values of 13.7–22.9% and 18.5–58.6%.

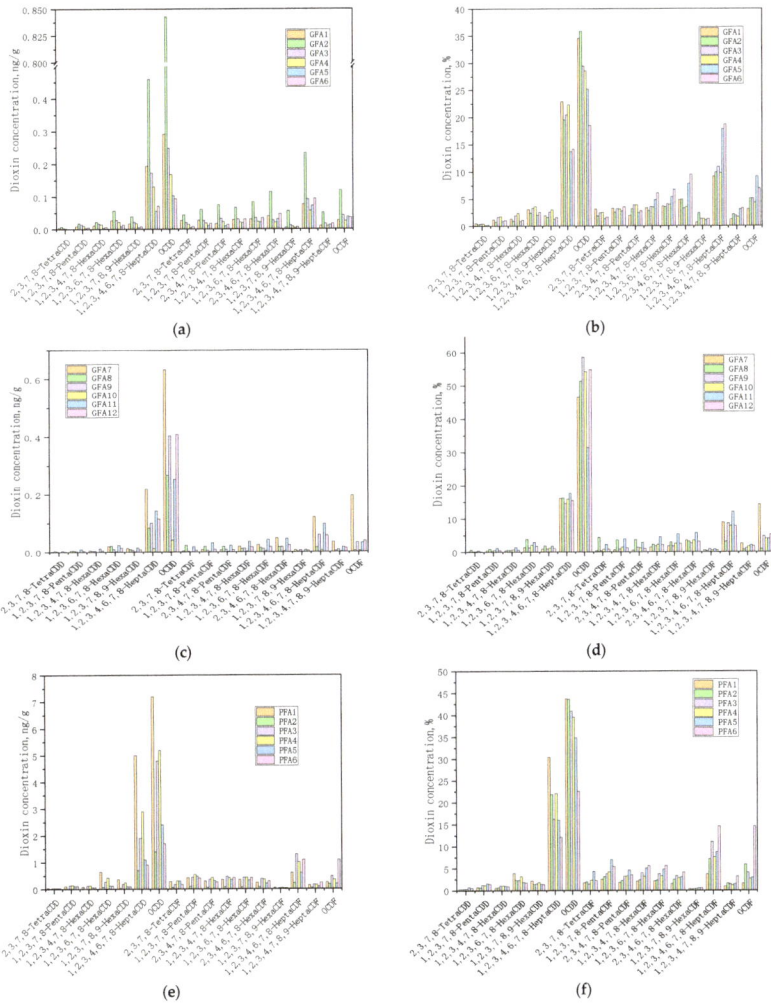

Figure 1. (**a**) Concentration distribution of PCDD/F in GFA1-GFA6; (**b**) percentage concentration distribution of PCDD/F in GFA1-GFA6; (**c**) concentration distribution of PCDD/F in GFA7-GFA12; (**d**) percentage concentration distribution of PCDD/F in GFA7-GFA12; (**e**) concentration distribution of PCDD/F in PFA1-PFA6; (**f**) percentage concentration distribution of PCDD/F in PFA1-PFA 6.

In contrast, the values in the previous study were 12.1–30.4% and 22.6–43.7%. The highest levels of PCDFs in this study were for 1,2,3,4,6,7,8-HpCDF (3.1–18.7%) and OCDF

(0.9–14.3%), which exhibited a content of 3.7–14.6% and 1.6–14.6% in six types of PFA samples (from the previous study). Other components exhibited a lower content of and a small difference in PCDFs, especially 2,3,7,8-TCDF, 1,2,3,7,8-PeCDF, 2,3,4,7,8-PeCDF, 1,2,3,4,7,8-HxCDF, and 1,2,3,6,7,8-HxCDF. This was possibly due to an insufficient amount of synthetic precursors [20].

3.3. Distribution of 10 Dioxin Homologues in the Samples

As compared with the 17 types of toxic dioxins, the distribution regularity of the 10 types of homologues was more obvious. The total homologue content distribution in the 18 types of fly ash is shown in Figure 2a,c,e. The dioxin concentration in grate furnace fly ash in this study was in the range of 0.162–6.685 ng/g, with a mean value of 2.418 ng/g, and the dioxin concentration in PFA was in the range of 9.8–55 ng/g, with a mean value of 33.97 ng/g. As can be seen, the concentration of the two is very different. It can be observed in Figure 2b that the distribution of dioxin content in GFA1-6 was more balanced and slightly bimodal, while some samples exhibited a unimodal distribution.

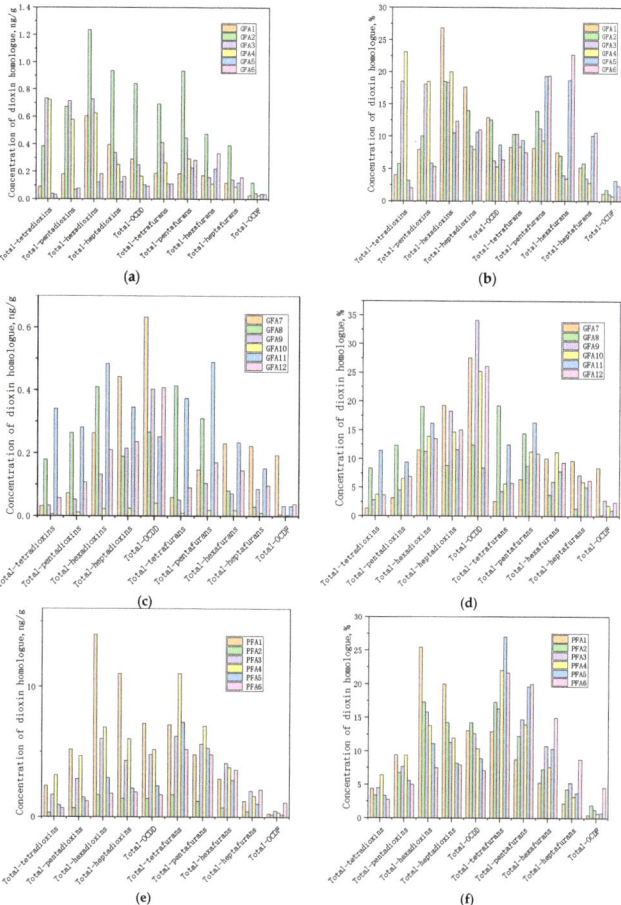

Figure 2. (a) Concentration distribution of 10 dioxin homologues in GFA1-GFA6; (b) percentage concentration distribution of 10 dioxin homologues in GFA1-GFA6; (c) concentration distribution of 10 dioxin homologues in GFA7-GFA12; (d) percentage concentration distribution of 10 dioxin homologues in GFA7-GFA12; (e) concentration distribution of 10 dioxin homologues in PFA1-PFA6; (f) percentage concentration distribution of 10 dioxin homologues in PFA1-PFA6.

The contents of Total-pentadioxins, Total-hexadioxins, and Total-heptadioxins in OCDD and Total-pentafurans and Total-hexafurans in OCDF were all higher than 10%. In Figure 2d, GFA6-12 exhibited a normal distribution characterized as high in the middle and low on both sides. OCDD (8.4–34.2%) was the highest and the others were marginally lower. In Figure 2f, the samples exhibited a distinct bimodal distribution, with Total-hexadioxins and Total-tetrafurans having the highest concentrations in PCDD and PCDF, with contents of 7.5–25.5% and 12.9–27%, respectively. In addition, the ratio of PCDD/PCDF in GFA ranged from 0.594 to 3.005, with an average of 1.756, and the PFA was in the range of 0.429–2.438, with a mean value of 1.164; thus, de novo synthesis dominated the trend in GFA [21,22].

3.4. Analysis of Dioxin Toxicity and Correlation of 17 Homologues

The analysis of the correlation between dioxins and toxic homologues was used to estimate the concentration of dioxins, which served as an important basis for distinguishing the distribution of dioxins. The data shown in Table 3 were obtained by using Excel to calculate the linear fit between 17 homologues and dioxin toxicity. In this study, the correlation coefficients of 1,2,3,7,8-PentaCDD, 1,2,3,7,8-PentaCDF, and 2,3,4,7,8-PentaCDF were very high, i.e., 0.9819, 0.9741, and 0.9735, respectively, as shown in Figure 3. They were used as indicator homologues, and the toxicity of dioxins was estimated by measuring these types of dioxins solely. This method was used in previous studies [16,23].

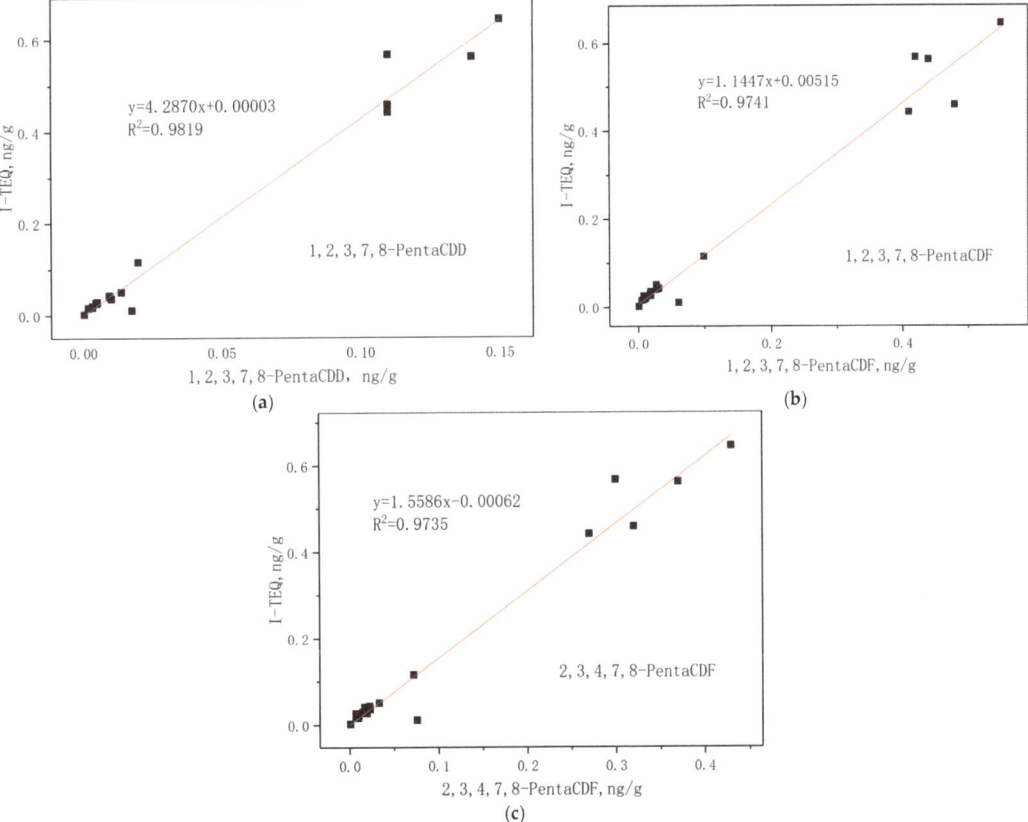

Figure 3. (a) Linear fit function for 1,2,3,7,8-PentaCDD; (b) linear fitting function for 1,2,3,7,8-PentaCDF; (c) linear fitting function for 2,3,4,7,8-PentaCDF.

Table 3. Relationship between dioxin homologues and dioxin toxicity.

PCDD/Fs Congeners	18 Sets of Data	
	Regression Equation	R^2
2,3,7,8-TetraCDD	12.851x − 0.00187	0.9472
1,2,3,7,8-PentaCDD	4.2870x + 0.000031	0.9819
1,2,3,4,7,8-HexaCDD	5.0952x − 0.00050	0.9263
1,2,3,6,7,8-HexaCDD	1.1691x + 0.04904	0.7288
1,2,3,7,8,9-HexaCDD	2.1281x + 0.03297	0.8007
1,2,3,4,6,7,8-HeptaCDD	0.1510x + 0.05254	0.6969
OCDD	0.1016x + 0.02329	0.8252
2,3,7,8-TetraCDF	2.0379x + 0.00806	0.9086
1,2,3,7,8-PentaCDF	1.1447x + 0.00515	0.9741
2,3,4,7,8-PentaCDF	1.5586x − 0.00062	0.9735
1,2,3,4,7,8-HexaCDF	1.3175x − 0.0014	0.9554
1,2,3,6,7,8-HexaCDF	1.3584x − 0.00732	0.9571
2,3,4,6,7,8-HexaCDF	1.7225x − 0.02329	0.9104
1,2,3,7,8,9-HexaCDF	8.9856x − 0.00372	0.6124
1,2,3,4,6,7,8-HeptaCDF	0.5239x + 0.00321	0.8475
1,2,3,4,7,8,9-HeptaCDF	2.9356x − 0.00110	0.8147
OCDF	0.5866x + 0.06933	0.4402

3.5. Distribution of Heavy Metals in Fly Ash Leaching Fluid

Heavy metals in fly ash are usually the focus of pollutants, and the content of heavy metals in different types of fly ash varies greatly [24]. Table 4 shows the results regarding the detection of heavy metals in various fly ash samples. Among the heavy metal indices, for Be, Cr, Ni, Cu, Pb, Zn, Se, Ba, Cr^{6+}, and As, the leaching concentration of fly ash was lower than the national standard. In GFA7, Hg exceeded the standard by twofold, and Cd exceeded the national standard by an order of magnitude. In GFA3, GFA10, and GFA11, Pb levels were very close to the standard value, which shows that the fly ash chelating agent is not very effective at stabilizing Pb in the fly ash tested. The concentration of Pb with low boiling point is higher in fly ash [25]. Pb also poses a high environmental risk [26], and the newly developed chelating agent should be studied in a targeted manner [27].

Table 4. Heavy metal leaching concentration of fly ash (mg/L).

Num	Be	Cr	Ni	Cu	Zn	Se	Cd	Ba	Pb	Cr^{6+}	Hg	As
GFA3	ND [1]	0.37	ND	ND	0.23	0.0022	ND	4.21	0.15	0.128	ND	0.0048
GFA4	ND	0.5	ND	ND	0.23	0.0031	ND	3.25	ND	0.124	ND	0.0065
GFA5	ND	1.06	0.101	0.017	0.02	0.019	ND	0.506	ND	1.05	0.0001	0.003
GFA6	ND	1.1	0.096	0.016	0.03	0.018	ND	0.508	ND	1.1	0.0001	0.004
GFA7	ND	ND	0.23	0.15	ND	ND	4.21	ND	0.0048	0.37	0.128	0.0022
GFA10	0.004	0.154	0.282	0.16	4	0.013	0.039	0.86	0.204	ND	0.0002	0.127
GFA11	ND	0.065	0.108	0.024	0.14	0.068	0.002	1.46	0.186	ND	0.0004	0.009
GFA12	ND	0.03	0.064	0.018	0.24	0.029	0.0029	0.68	0.0050	ND	0.0002	ND
GB [2]	0.02	4.5	0.5	40	100	0.1	0.15	25	0.25	1.5	0.05	0.3

[1] refers to undetected, [2] refers to the current general administration of quality supervision, inspection and quarantine 16889-2008: Standard for pollution control on the landfill site of municipal solid waste.

4. Conclusions

To the best of our knowledge, this is the first study to comprehensively analyze fly ash after chelating agent treatment. The fly ash tested was sampled from a large proportion of power plants in China. Grading the pollution associated with fly ash is important for understanding the current levels of fly ash dioxins in landfills. All power plants were shown to be well below the national standard for dioxin emissions, exhibiting a range of 0.002–0.051 ngI-TEQ/g, with an average of 0.027ngI-TEQ/g, and even below 50 ng/kg, which is the fly ash resource utilization standard. Among the 17 toxic dioxins, the main contributors to PCDD/Fs were consistent, but their relative contents were different. Among the 10 dioxin homologues, the dioxin content distribution in GFA was relatively balanced, showing slightly bimodal, unipolar, and normal characteristics. In PFA, the samples exhibited a distinct bimodal distribution. In a linear fit, the correlation coefficients for 1,2,3,7,8-PentaCDD, 1,2,3,7,8-PentaCDF, and 2,3,4,7,8-PentaCDF were very high, being 0.9819, 0.9741, and 0.9735, respectively. In this heavy metals analysis, the fly ash was treated with a chelating agent, and the heavy metal content was shown to be lower than obtained with the national standard chelating agent, except for one type of fly ash. However, a good chelating effect was not observed for Pb. In addition, all the fly ash waste that was sampled and analyzed for this study is currently landfilled, consuming valuable space and resources. The problem of eradicating toxic substances has become the largest obstacle to the application of fly ash [28].

The fly ash in this study was taken solely from ash hoppers. In waste incineration plants, fly ash spreads to the surrounding environment, harming the health of on-site workers and surrounding residents. The diffused fly ash may exhibit different distributions of dioxins and heavy metals. In addition, different plants may have different tolerances to pollutants in fly ash, which may depend on the absorption of fly ash in the air and soil. This will be the focus of our next stage of research.

Author Contributions: Conceptualization, S.P. and S.L.; methodology, S.P.; software, Q.Y.; validation, Q.Y.; formal analysis, Y.P.; investigation, Y.L. and Z.W.; resources, X.L.; data curation, Y.P.; writing—original draft preparation, Q.Y.; writing—review and editing, W.C. and C.J.; visualization, Q.Y.; supervision, S.L., X.L., Y.L. and Z.W.; project administration, W.C. and C.J.; funding acquisition, W.C. All authors have read and agreed to the published version of the manuscript.

Funding: This research and the APC was funded by the "Pioneer" and "Leading Goose" R&D Program of Zhejiang (2022C03056).

Data Availability Statement: Not applicable.

Conflicts of Interest: The authors declare that they have no known competing financial interests or personal relationships that could have appeared to influence the work reported in this paper.

References

1. Song, J.; Sun, Y.; Jin, L. PESTEL analysis of the development of the waste-to-energy incineration industry in China. *Renew. Sustain. Energ. Rev.* **2017**, *80*, 276–289. [CrossRef]
2. Zhao, F.; Bian, R.; Zhang, T.; Fang, X.; Chai, X.; Xin, M.; Li, W.; Sun, Y.; Yuan, L.; Chen, J.; et al. Characteristics of polychlorinated dibenzodioxins/dibenzofurans from a full-scale municipal solid waste (MSW) incinerator in China by MSW classification. *Process Saf. Environ.* **2022**, *161*, 50–57. [CrossRef]
3. Yearbook, C.S. *National Bureau of Statistics of China*; China Statistics Press: Beijing, China, 2021.
4. Zhu, J.; Wei, Z.; Luo, Z.; Yu, L.; Yin, K. Phase changes during various treatment processes for incineration bottom ash from municipal solid wastes: A review in the application-environment nexus. *Environ. Pollut.* **2021**, *287*, 117618. [CrossRef]
5. Allegrini, E.; Butera, S.; Kosson, D.S.; Van Zomeren, A.; Van der Sloot, H.A.; Astrup, T.F. Life cycle assessment and residue leaching: The importance of parameter, scenario and leaching data selection. *Waste Manag.* **2015**, *38*, 474–485. [CrossRef] [PubMed]
6. Kulkarni, P.S.; Crespo, J.G.; Afonso, C.A.M. Dioxins sources and current remediation technologies—A review. *Environ. Int.* **2008**, *34*, 139–153. [CrossRef]
7. Buekens, A.; Huang, H. Comparative evaluation of techniques for controlling the formation and emission of chlorinated dioxins/furans in municipal waste incineration. *J. Hazard. Mater.* **1998**, *62*, 1–33. [CrossRef]
8. Xue, Y.; Liu, X. Detoxification, solidification and recycling of municipal solid waste incineration fly ash: A review. *Chem. Eng. J.* **2021**, *420*, 1303493. [CrossRef]
9. Qiu, Q.; Chen, Q.; Jiang, X.; Lv, G.; Chen, Z.; Lu, S.; Ni, M.; Yan, J.; Lin, X.; Song, H.; et al. Improving microwave-assisted hydrothermal degradation of PCDD/Fs in fly ash with added Na2HPO4 and water-washing pretreatment. *Chemosphere* **2019**, *220*, 1118–1125. [CrossRef]
10. Liu, H.; Zeng, T.; Wei, G.; Zhang, R.; Liu, F.; Wang, H. Comparison of Dioxin Destruction in the Fly Ash and Froths under Microwave Irradiation. *Aerosol Air Qual. Res.* **2019**, *19*, 925–936. [CrossRef]
11. Yap, H.C.; Pang, Y.L.; Lim, S.; Abdullah, A.Z.; Ong, H.C.; Wu, C.H. A comprehensive review on state-of-the-art photo-, sono-, and sonophotocatalytic treatments to degrade emerging contaminants. *Int. J. Environ. Sci. Technol.* **2019**, *16*, 601–628. [CrossRef]

12. Li, W.; Gu, K.; Yu, Q.; Sun, Y.; Wang, Y.; Xin, M.; Bian, R.; Wang, H.; Wang, Y.; Zhang, D. Leaching behavior and environmental risk assessment of toxic metals in municipal solid waste incineration fly ash exposed to mature landfill leachate environment. *Waste Manag.* **2021**, *120*, 68–75. [CrossRef] [PubMed]
13. Takasuga, T.; Makino, T.; Tsubota, K.; Takeda, N. Formation of dioxins (PCDDs/PCDFs) by dioxin-free fly ash as a catalyst and relation with several chlorine-sources. *Chemosphere* **2000**, *40*, 1003–1007. [CrossRef]
14. He, H.; Lu, S.; Peng, Y.; Tang, M.; Zhan, M.; Lu, S.; Xu, L.; Zhong, W.; Xu, L. Emission characteristics of dioxins during iron ore Co-sintering with municipal solid waste incinerator fly ash in a sintering pot. *Chemosphere* **2022**, *287*, 131884. [CrossRef] [PubMed]
15. Chen, T.; Yan, J.H.; Lu, S.Y.; Li, X.D.; Gu, Y.L.; Dai, H.F.; Ni, M.J.; Cen, K.F. Characteristic of polychlorinated dibenzo-p-dioxins and dibenzofurans in fly ash from incinerators in china. *J. Hazard. Mater.* **2008**, *150*, 510–514. [CrossRef] [PubMed]
16. Pan, Y.; Yang, L.; Zhou, J.; Liu, J.; Qian, G.; Ohtsuka, N.; Motegi, M.; Oh, K.; Hosono, S. Characteristics of dioxins content in fly ash from municipal solid waste incinerators in China. *Chemosphere* **2013**, *92*, 765–771. [CrossRef]
17. Ni, Y.; Zhang, H.; Fan, S.; Zhang, X.; Zhang, Q.; Chen, J. Emissions of PCDD/Fs from municipal solid waste incinerators in China. *Chemosphere* **2009**, *75*, 1153–1158. [CrossRef]
18. Chen, Z.; Mao, Q.; Lu, S.; Buekens, A.; Xu, S.; Wang, X.; Yan, J. Dioxins degradation and reformation during mechanochemical treatment. *Chemosphere* **2017**, *180*, 130–140. [CrossRef]
19. Chang, Y.; Fan, W.; Dai, W.; Hsi, H.; Wu, C.; Chen, C. Characteristics of PCDD/F content in fly ash discharged from municipal solid waste incinerators. *J. Hazard. Mater.* **2011**, *192*, 521–529. [CrossRef]
20. Wikstrom, E.; Tysklind, M.; Marklund, S. Influence of variation in combustion conditions on the primary formation of chlorinated organic micropollutants during municipal solid waste combustion. *Environ. Sci. Technol.* **1999**, *33*, 4263–4269. [CrossRef]
21. Everaert, K.; Baeyens, J. The formation and emission of dioxins in large scale thermal processes. *Chemosphere* **2002**, *46*, 439–448. [CrossRef]
22. Zhang, G.; Huang, X.; Liao, W.; Kang, S.; Ren, M.; Hai, J. Measurement of Dioxin Emissions from a Small-Scale Waste Incinerator in the Absence of Air Pollution Controls. *Int. J. Environ. Res. Public Health* **2019**, *16*, 1267. [CrossRef] [PubMed]
23. Minomo, K.; Ohtsuka, N.; Nojiri, K.; Hosono, S.; Kawamura, K. A Simplified Determination Method of Dioxin Toxic Equivalent (TEQ) by Single GC/MS Measurement of Five Indicative Congeners. *Anal. Sci.* **2011**, *27*, 421–426. [CrossRef] [PubMed]
24. Li, R.; Nie, Y.; Wang, L.; Li, A.; Chi, Y.; Cen, K. Trace Pollutant in MSW Incinerator Fly Ash:Heavy Metals and Dioxins. *J. Combust. Sci. Technol.* **2004**, *10*, 479–483.
25. Tanigaki, N.; Fujinaga, Y.; Kajiyama, H.; Ishida, Y. Operating and environmental performances of commercial-scale waste gasification and melting technology. *Waste Manag. Res.* **2013**, *31*, 1118–1124. [CrossRef]
26. Xiong, Y.; Zhu, F.; Zhao, L.; Jiang, H.; Zhang, Z. Heavy metal speciation in various types of fly ash from municipal solid waste incinerator. *J. Mater. Cycles Waste* **2014**, *16*, 608–615. [CrossRef]
27. Zhu, Z.; Guo, Y.; Zhao, Y.; Zhou, T. A novel waste-recycled chelating agent for the stabilization of lead in municipal solid waste incineration fly ash: Preparation, feasibility, and mechanism analysis. *J. Hazard. Mater.* **2022**, *427*, 127914. [CrossRef]
28. Chen, D.; Zhang, Y.; Xu, Y.; Nie, Q.; Yang, Z.; Sheng, W.; Qian, G. Municipal solid waste incineration residues recycled for typical construction materials—A review. *Rsc. Adv.* **2022**, *12*, 6279–6291. [CrossRef]

Article

Evaluation of the Immobilization of Fly Ash from the Incineration of Municipal Waste in Cement Mortar Incorporating Nanomaterials—A Case Study

Monika Czop [1,*], Beata Łaźniewska-Piekarczyk [2] and Małgorzata Kajda-Szcześniak [1]

[1] Department of Technologies and Installations for Waste Management, Faculty of Energy and Environmental Engineering, The Silesian University of Technology, Konarskiego 18, 44-100 Gliwice, Poland
[2] Department of Building Processes and Building Physics, Faculty of Civil Engineering, The Silesian University of Technology, Akademicka 5, 44-100 Gliwice, Poland
* Correspondence: monika.czop@polsl.pl; Tel.: +48-32-237-21-04

Citation: Czop, M.; Łaźniewska-Piekarczyk, B.; Kajda-Szcześniak, M. Evaluation of the Immobilization of Fly Ash from the Incineration of Municipal Waste in Cement Mortar Incorporating Nanomaterials—A Case Study. *Energies* **2022**, *15*, 9050. https://doi.org/10.3390/en15239050

Academic Editors: Robert Oleniacz and Katarzyna Grzesik

Received: 8 November 2022
Accepted: 26 November 2022
Published: 29 November 2022

Publisher's Note: MDPI stays neutral with regard to jurisdictional claims in published maps and institutional affiliations.

Copyright: © 2022 by the authors. Licensee MDPI, Basel, Switzerland. This article is an open access article distributed under the terms and conditions of the Creative Commons Attribution (CC BY) license (https://creativecommons.org/licenses/by/4.0/).

Abstract: Fly ash generated in the process of combustion of municipal waste is classified as hazardous waste. Its management today has become a significant problem. One of the methods of safe management of such ash may be using it for the production of concrete as a partial replacement for cement. Using immobilization, the number of hazardous compounds could be limited so that the obtained new material would be safe for the natural environment. Recovery of byproducts—in this case, fly ash—complies with the business models applied in the production cycle in the circular economy model. Such a solution may result in saving energy, limiting CO_2 emissions, reducing the use of natural resources, and management of dangerous waste. It should be added that concretes with the addition of hazardous waste would be used for industrial purposes according to the binding legal regulations. This article presents the influence of the addition of fly ash on the selected mechanical properties of concrete. Fly ash from the incineration of municipal waste was used as a partial replacement of CEM I concrete at amounts of 4%, 8%, and 18% of its mass. The compressive strength and flexural strength of such concretes were tested after 28 days of concrete curing. This article also presents the tests of the leachability of contaminants from fly ash and concretes produced with Portland cement CEM I. The test results confirm that immobilization is an effective process that limits the amount of contamination in the water extract. Zinc, lead, and chrome were almost completely immobilized by the C-S-H (calcium silicate hydrate) concrete phase, with their immobilization degree exceeding 99%. Chloride content also underwent immobilization at a similar level of 99%. The sulfates were immobilized at the level of 96%. The subject matter discussed in this article is essential because, to protect the natural environment and, thus, reduce the use of natural resources, it is increasingly necessary to reuse raw materials—not natural, but recycled from the industry. Waste often contains hazardous compounds. A proposal for their safe disposal is their immobilization in a cement matrix. An important aspect is reducing leachability from concrete as much as possible, e.g., using nanomaterials. The effectiveness of reducing the leachability of hazardous compounds with the proposed method was checked in this study.

Keywords: combustion; fly ash; recovery; immobilization; leaching

1. Introduction

The constant development of the economy and the worldwide increase in consumption are reflected in the production of an enormous amount of municipal waste. According to the provisions of Directive 2008/98/EC [1] and the national waste law [2], each action related to waste must comply with the binding hierarchy of dealing with waste. According to the listed legal acts, first of all, it is recommended to prepare the wastes for their reuse, after which they should undergo recycling or other recovery processes, including energy recovery. The last element of the waste management hierarchy is their neutralization, e.g., by storage [1,2].

According to Statistics Poland's data, in 2021 in Poland, 13.6 M tons of municipal waste was collected. This represents an increase of about 4.2% compared to 2020 [3]. In 2021, 60% of the stream of collected municipal waste was directed for recovery (82,070 thousand tons), and 40% was neutralized. It needs to be underlined that in 2021 a stream of mixed fractions constituted 60% of all of the collected municipal waste, accounting for 8234 thousand tons [3]. This represents a decrease of 2% compared to 2020. Depositing mixed municipal wastes in landfills is no longer reasonable. An alternative is thermal processing with energy recovery. The idea of constructing an incineration plant is very popular, as it at least partially solves the problems that many local governments face. In order to decrease the environmental issues and follow the idea of a circular economy (i.e., the concept of rationally reusing resources and limiting the negative impacts of manufactured products on the environment), it is worth identifying other methods of management of combustion byproducts such as fly ash [4–10]. Fly ash generated as a result of incineration may be used in the construction industry, saving natural resources. It needs to be emphasized that the construction materials produced using fly ash from incineration plants are most frequently characterized by low strength and, therefore, they are used mainly for the construction of local roads. Irrespective of where they are applied, the final product cannot be harmful to the environment [11]. In the scientific literature [12–19], there is little information on how fly ash from municipal waste incineration plants affects the properties of cement mortars, making the subject discussed in this paper a novelty. However, the use of fly ash in construction has been recognized for a long time now [20–27]—for example, considering fly ash generated as a side product during the combustion of hard coal or biomass. Because of its properties, fly ash is used as a mineral additive to cement; as a matrix for sand formation, concrete admixtures, bituminous masses, and ceramic tiles; in geotechnics; in the foundations of road pavement; and for soil stabilization [22,25].

According to the data found in the literature, the immobilization process is usually applied to hazardous wastes such as industrial dust and sludge, galvanic waste, gravels and ashes from thermal processes (e.g., in the iron and steel industry), nonferrous metals from municipal waste incineration plants, waste sediments, or dusts and sludges from gas purification processes [6,11,28–33]. In many cases, the basic management method of such hazardous wastes should be the immobilization of the waste, due to its low costs and the possibility of managing a wide range of wastes. Solidified waste is not harmful to the environment and may be applied in industry [11,28]. Solidification/stabilization methods may be divided into six groups depending on the main components and processes applied: cement-based, lime-based, based on thermoplastic processes, based on organic polymers, based on encapsulation, or vitrification processes. The abovementioned process groups vary in their application, costs, and requirements for the initial processing of the waste. However, all of them aim at modifying the physicochemical properties of the wastes in such a way as to limit the migration of the contaminants to the environment, produce homogenous concrete matrixes suitable for their reuse, and facilitate the transport and disposal of wastes to landfills [34–36]. This article analyzes fly ash obtained from the seasonal removal of ash from boilers and water heaters. The purpose of this work was to test the elemental composition of the ash in light of its impact on the environment and to determine the possibilities of using it for the production of modern construction materials. This idea fits into the concept of the circular economy (CE).

The tests adhered to the CE business model, which is based on the recovery of the side products. This model consists of actions where residues or secondary products of one process become inputs for another process. The actual management method of fly ash from the incineration of municipal waste is storage in landfills, where the potential raw material is irreversibly lost. The undertaken actions aim at extending the life cycle of the fly ash. It is estimated that the extension of the fly ash life cycle will result in a decrease in the use of natural resources, which translates into economic and financial savings. Today, the possibility of using fly ash from the incineration of municipal waste in the construction industry is under investigation. Fly ash could be a valuable material for the production of cement, concrete, and precast elements for industrial use.

2. The Analyzed Installation of Thermal Processing of Municipal Waste—Case Study

The discussed municipal waste incineration plant (MSWI) is located in Poland in the Lesser Poland Voivodeship—specifically in the southeast part of the city of Cracow in District XVIII (Nowa Huta). The analyzed MSWI mainly transforms mixed municipal waste. Additionally, residues from the mechanical processing of municipal fractions and large construction waste products are combusted [4]. The waste comes from the area of Cracow City Commune. The yearly capacity of the plant is 220,000 Mg, and the calorific value of the waste is 8.8 MJ/kg. The thermal power of the plant is 35 MWt, and the electrical power is 10.7 MWe [4]. On two parallel lines, 700 tons of waste is combusted within 24 h. The waste from incineration processes includes boiler dust, fly ash, and solid residues from the purification of the exhaust fumes. It is estimated that the residues from the incineration of municipal waste constitute approx. 25% of the input stream. Hazardous waste products—i.e., boiler dust, fly ash, and solid residues from the purification of the exhaust fumes—are transferred with a pneumatic transporter to silos adapted to cistern loading. This makes it possible to transport them for further processing and management in other specialist plants or for deep storage in former salt mine pits (e.g., in Germany).

Figure 1 presents a simplified scheme of the analyzed municipal solid waste incineration plant.

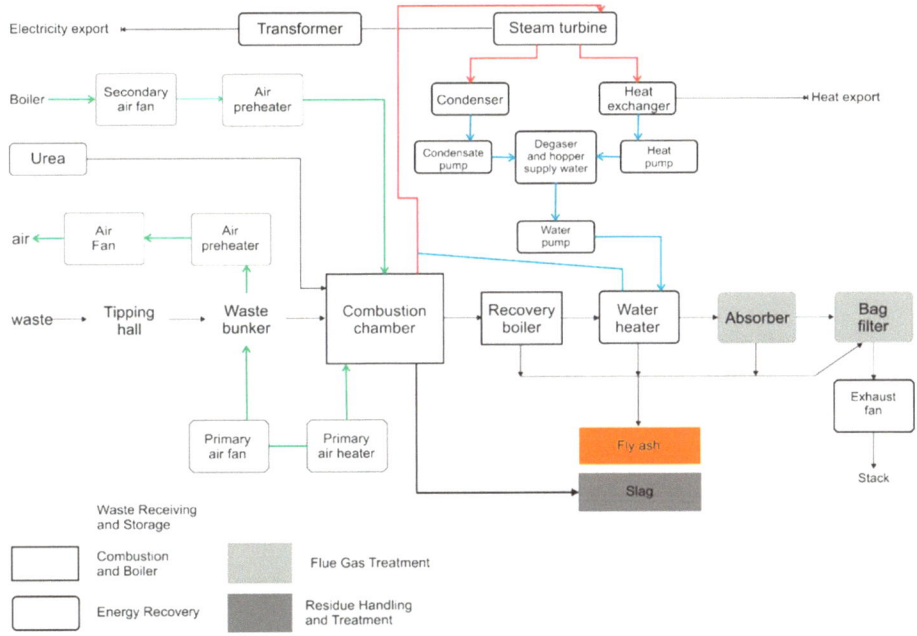

Figure 1. Simplified scheme of the analyzed municipal solid waste (MSW) incinerator.

3. Materials

The material that was the basis for the tests was fly ash (FA; Figure 2) generated in the process of incineration of municipal waste in grid furnaces. It consisted mainly of non-flammable substances (e.g., silicates, aluminum, and iron oxides that are insoluble in water). According to European Waste Codes [37,38], it has a code of 19 01 13*—fly ash containing hazardous substances. This is hazardous waste and may have the following characteristics: irritant, dangerous, toxic, allergenic, and ecotoxic. However, to a small degree, it has some features of a flammable fraction—unburned coal particles (Figure 2a).

Figure 2. The tested fly ash (FA): (**a**) draw; (**b**) mechanically crushed.

Figure 3c presents the loss on ignition (LOI) determined according to the standard PN-EN 15935:2013-02 [39]. The tested fly ash was incinerated until constant mass at temperatures of 600 °C and 950 °C in a laboratory muffle furnace. Fly ash can be divided into three categories based on the loss on ignition determined at a temperature of 950 °C, according to the standard PN-EN 450-1:2012 [40]: category A (LOI ≤ 5%), category B (LOI ≤ 7%), and category C (LOI ≤ 9%) (Figure 3a). Based on the performed analyses, the tested fly ash could be classified as category B. Furthermore, with the loss on ignition determined at a temperature of 600 °C, it was noted that the tested fly ash met the requirements for waste other than hazardous (LOI ≤ 8%) or neutral waste (LOI ≤ 10%) that are accepted at the landfill [41] (Order of the Minister of Economy 2015) (Figure 3b).

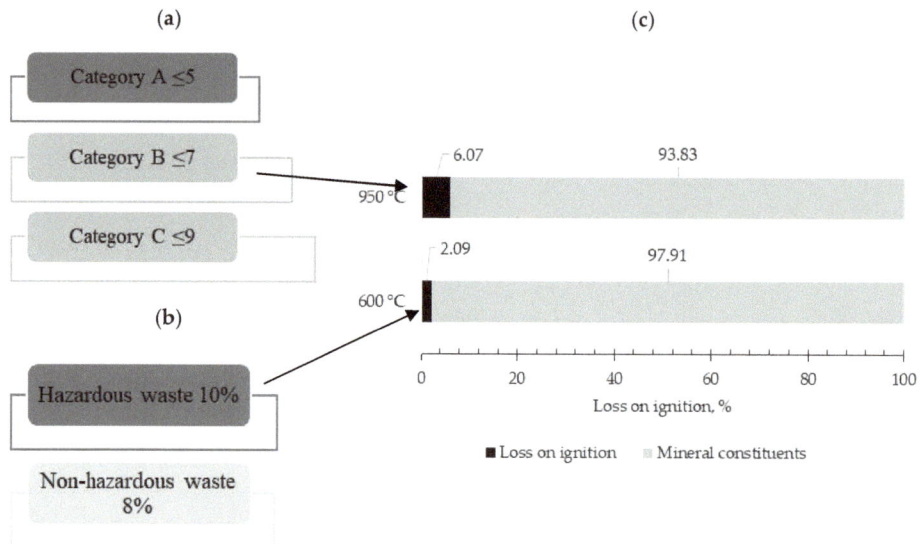

Figure 3. Loss on ignition (LOI): (**a**) requirements; (**b**) other criteria for waste acceptable at landfills; (**c**) tested fly ash.

At the beginning of the research, an analysis of oxide and heavy metal contents was carried out. The obtained results (Tables 1 and 2, respectively) were compared with the requirements concerning, among others, the chemical properties of fly ash used as a type II additive for the production of concrete [40] (PN-EN 450-1:2012). The main phase components of fly ash from the municipal solid waste incineration plant were CaO and SiO_2.

Table 1. Contents of oxides (%) in the tested fly ash.

Parameter	Symbol	Fly Ash		Requirements for Fly Ash for the Production of Concretes (PN-EN 450-1:2012) [40]
Silicon dioxide	SiO_2	29.50		
Iron (III) oxide	Fe_2O_3	2.81	41.60	$\Sigma SiO_2, Al_2O_3$ i $Fe_2O_3 \geq 70$
Aluminum oxide	Al_2O_3	9.29		
Manganese (II, III) oxide	Mn_3O_4	0.12		nr **
Titanium dioxide	TiO_2	2.23		nr **
Calcium oxide	CaO	30.10		nr **
Magnesium oxide	MgO	2.60		≤ 4.0
Sulfur trioxide	SO_3	8.71		≤ 3.0
Phosphorus pentoxide	P_2O_5	1.77		≤ 5.0
Sodium oxide	Na_2O	2.67		nr **
Potassium oxide	K_2O	1.95		nr **
Barium oxide	BaO	0.22		nr **
Strontium oxide	SrO	0.06		nr **

** nr—no requirements.

Table 2. Heavy metal concentrations, expressed in mg/kg.

Parameter	Symbol	Fly Ash
Zinc	Zn	7242.0
Copper	Cu	325.0
Lead	Pb	586.0
Nickel	Ni	113.0
Chrome	Cr	334.0
Cadmium	Cd	23.3
Arsenic	As	7.3
Vanadium	V	39.1
Thallium	Tl	<1.0
Mercury	Hg	0.02

The concentrations of heavy metals in the dry mass of tested FA were high, and the sequence was as follows: Zn > Pb > Cr > Cu > Ni > V > Cd > As > Tl > Hg. The highest value was reported for zinc, which reached 7242.0 mg/kg, while the lowest was recorded for thallium (<1.0 mg/kg).

4. Methods

The testing procedure was planned and carried out in such a way as to determine the characteristics of the fly ash from the MSWI plant with respect to the physical and chemical properties that are important in the context of the use of fly ash as a partial replacement for cement. The impact on the environment was also taken into account. The testing procedure included eight stages:

- Testing the physicochemical properties of the fly ash;
- Preparation of the aqueous extract with fly ash considering the impact on the environment;
- Designing and preparing mortars with the 4% and 18% addition of fly ash;
- Testing the flexural and compressive strengths of the mortars (beams 40 × 40 × 160 mm) produced with 4% and 18% addition of FA as compared to the reference sample;
- Preparation of aqueous extracts with crushed mortars after 28 days of curing, and executing chemical tests, with an evaluation of the impact on the environment;
- Designing and preparing mortars with 4% and 8% addition of fly ash modified with nanomaterials;
- Testing the flexural and compressive strengths of the designed cement mortars (beams 40 × 40 × 160 mm) with 4% and 8% addition of FA modified with nanomaterials as compared to the reference sample;

- Preparation of aqueous extracts with crushed mortars after 28 days of curing, and executing chemical tests, with an evaluation of the impact on the environment.

4.1. The Procedure for the Preparation of the Aqueous Extract with Fly Ash

The aqueous extract was produced according to the standard PN-EN 12457-2:2006 [42]. From the 2 kg of sample ash, a representative laboratory sample was prepared. For the purpose of the analysis, the tested ash was sieved through the screen of a 2 mm mesh. From this sample, an aqueous extract was prepared with a liquid/solid ratio (L/S) of 10 L/kg. The elution water was distilled water with pH 7.4 and electrical conductivity of 61.18 µS/cm. The prepared samples were shaken in a laboratory shaker for 24 h, and the obtained extracts were left for 15 min for decantation of solid particles, followed by filtering. The pH was determined using an Elmetron CPC-501 device (PN-EN ISO 10523:2012, PN-EN 27888:1999) [43,44]. The analysis of the aqueous extracts of fly ash included a number of specifications. The content of chlorides was determined via the Mohr method with the use of silver nitrate as a titration agent and potassium chromate as an indicator (PN-ISO 9297:1994) [45]. Sulfates (VI) (SO_4^{2-}) were determined via a gravimetric method with barium chloride (PN-ISO 9280:2002) [46]. The contents of sodium, calcium, potassium, lithium, and barium in the aqueous extracts from fly ash were determined via flame emission spectrometry (PN-ISO 9964-3:1994) [47]. The phosphorus content was determined as described in [48]. In order to evaluate the heavy metal composition (Zn, Cu, Pb, Cd, Cr, Co, Fe, Ni) in the aqueous extract, inductively coupled atomic absorption spectroscopy (AAS) was performed using GBC's AVANTA PM apparatus.

4.2. Composition and Methodology of Preparation of Cement Mortars with the Addition of Fly Ash

The subject of the test was cement mortars with the addition of fly ash from the municipal waste incineration plant, with and without modifications with nanomaterials. The mortars were produced with Portland cement CEM I 52.5R (ÓRAŻDŻE CEMENT S.A., Poland) meeting the requirements of PN-EN 197-1 [49], and a standardized sand of fraction 0 ÷ 2 mm, compliant with PN-EN 196-1 [50]. Four cement mortars were prepared: CEM-I reference mortar; CEM I+4% FA—mortar with 4% fly ash from the incineration plant; CEM I + 4% FA + N—mortar with the addition of nanomaterial; and CEM I + 8% FA + N—mortar with 8% fly ash from the incineration plant and nanomaterial. The compositions of the mortars are specified in Table 3.

Table 3. Composition of the concrete mortars, expressed in grams.

Type of Waste	Symbol of Mortar	CEM I	Nano Al	Water	Sand Acc. (PN-EN 196-1) [50]
Reference sample from Portland cement 52.5R	CEM I	450	-	225	1350
CEM I 52.5R + 4% fly ash	CEM I + 4% FA	402.5	-	171	1350
CEM I 52.5R + 18% fly ash	CEM I + 18% FA	354.21	-	171	1350
CEM I 52.5R + 4% fly ash + nano-Al	CEM I + 4% FA + N	402.50	10.5	171	1350
CEM I 52.5R + 8% fly ash + nano-Al	CEM I + 8% FA + N	386.36	10.5	171	1350

Tests of the water demand of the cement grout and fly ash were carried out using a Vicata automatic device according to the standard PN-EN 196-3c:2016-12 [51]. Determination consisted of measuring the time required from mixing the grout components to the commencement and completion of the binding process. Subsequently, three beams of dimensions 4 × 4 × 16 cm and compliant with the standard PN-EN 196-1 [50] were made from each mortar. The samples were removed from their forms after 24 h and stored in water at a temperature of 20 °C ± 2 °C for 28 days. After 28 days of concrete curing, the flexural and compressive strengths of the samples were tested according to the standard PN-EN 196-1 [50].

4.3. The Procedure of Preparing Aqueous Extracts from the Crushed Cement Mortars with the Addition of Fly Ash

The aqueous extracts from the crushed mortar cement after 28 days of concrete curing were prepared according to the standard PN-EN 12457-4:2006 [52]. The mortars were crushed to a grain size of <10 mm and then shaken for 24 h, maintaining a liquid–solid ratio (L/S) of 10. The elution water was distilled water with pH 7.4 and conductivity of 61.18 µS/cm. After the completion of shaking, the obtained extracts were filtered. The analysis of the aqueous extracts from the crushed cement mortars was performed as described in Section 4.1.

5. Results and Discussion

5.1. The Evaluation of the Leachability of Hazardous Substances and Heavy Metals from the Fly Ash

Table 4 presents the leachability from fly ash of hazardous substances and heavy metals that may be a nuisance to the environment and negatively affect the properties of the concrete mix, potentially affecting the concrete's strength. The obtained results were compared to the binding national [41] and European [53] legal regulations. The tested fly ash was characterized by a strong alkaline reaction—above pH 12. The leachability of chlorides (Cl^-) and sulfates (SO_4^{2-}) from tested fly ash did not exceed the acceptable levels for depositing wastes other than hazardous and neutral wastes in landfills. Only the barium content (Ba) exceeded the permissible values for wastes other than hazardous and dangerous stored in landfills (by about 36%). The leachability of heavy metals in the tested fly ash could be considered to be low. The contents of Zn, Cr, and Pb did not exceed the permissible values, and in some cases (i.e., Cu, Cd, Ni) the contents were below the limit of quantitation.

Table 4. Leachability of hazardous substances and heavy metals from fly ash, expressed in mg/kg.

Parameter	Symbol	Fly Ash	Criteria for Landfills [41,53]	
			For Non-Hazardous Waste	For Hazardous Waste
pH	pH	12.9	min. 6	-
Chloride	Cl^-	352.51	15,000	25,000
Sulfate	SO_4^{2-}	12,350.23	20,000	50,000
Phosphate	PO_4^{3-}	<0.005	-	-
Potassium	K	49.82	-	-
Calcium	Ca	459.90	-	-
Lithium	Li	1.90	-	-
Sodium	Na	55.81	-	-
The sum of chloride and sulfate	($Cl^- + SO_4^{2-}$)	12,702.74	60,000	100,000
Barium	Ba	156.90	100	300
Zinc	Zn	9.55	50	200
Copper	Cu	<0.20	50	100
Lead	Pb	4.93	10	50
Cadmium	Cd	<0.05	1	5
Chrome	Cr	2.26	10	70
Cobalt	Co	0.50	-	-
Iron	Fe	0.40	-	-
Nickel	Ni	<0.40	10	40

5.2. Evaluation of the Degree of Immobilization of Contaminants from Cement Mortars with the Addition of Fly Ash

The cement batch with the addition of 4% fly ash showed higher water demand. The use of FA in the MSWI plant as a partial replacement for cement required the addition of more water when producing the concrete mix or concrete batch. Replacing 4% of the cement mass with FA resulted in minor shortening of the early binding time. The early binding time was 11 min shorter than in the reference batch, for which the manufacturer of cement 52.R guarantees an early binding time of 186 min. Figures 4 and 5 present the designed cement mortars with 4% and 18% fly ash from the incineration plant, respectively.

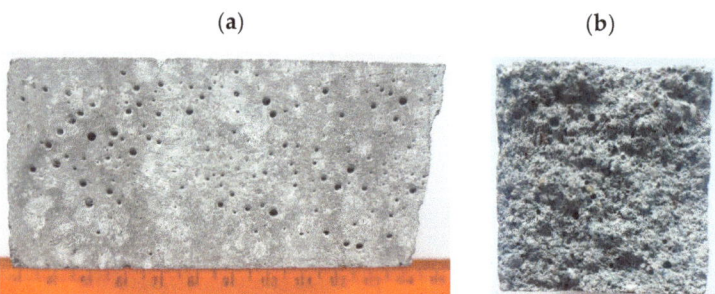

Figure 4. Cement mortar with the addition of 4% fly ash from the incineration plant (CEM I + 4% FA): (**a**) outer surface; (**b**) internal structural view.

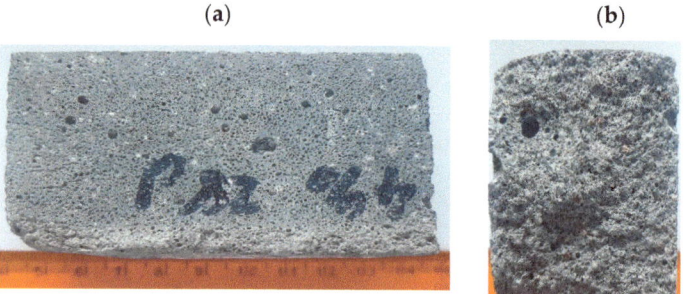

Figure 5. Cement mortar with the addition of 18% fly ash from the MSWI plant (CEM I + 18% FA): (**a**) outer surface; (**b**) internal structural view.

The results of the flexural and compressive strength tests of the cement mortars with 4% and 18% addition of fly ash from the incineration plant after 28 days of concrete curing are presented in Figure 6. The compressive strength of a beam with 4% addition of FA was 71.37 MPa, which was about 16 MPa less than the reference value. The flexural strength of the beam with 4% FA was about 8.85 MPa, which was about 0.75 MPa less than the reference value.

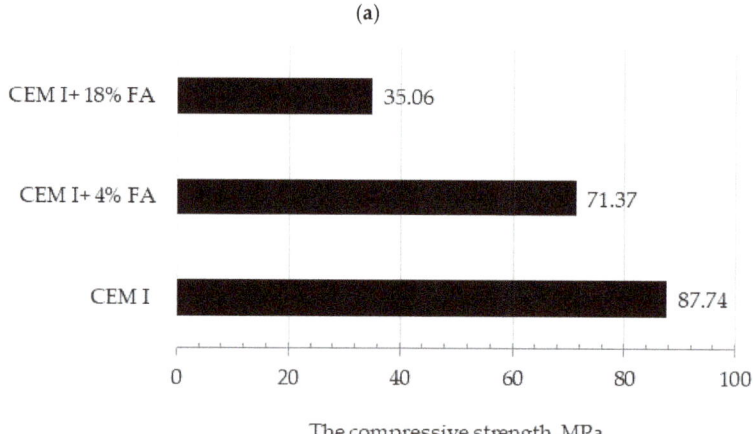

Figure 6. *Cont.*

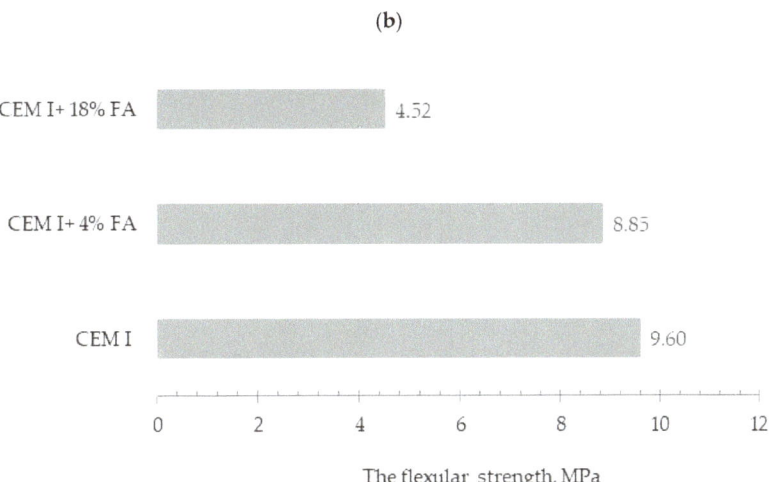

Figure 6. Strength of cement mortars with 4% and 18% addition of FA after 28 days of concrete curing: (a) compressive strength; (b) flexural strength.

In the case of concrete mortars with 18% fly ash, it was noted that the compressive strength and flexural strength were lower than in the reference sample, by about 52.68 MPa and 5.08%, respectively. Moreover, a swelling effect was observed (Figure 7). This is an undesirable effect because it causes volume changes in the concrete, resulting in cracking. Therefore, such material is useless for the construction industry. It was noted that 4% content of fly ash in the concrete mass is a limit value that does not negatively affect the mechanical properties of cement mortars. The use of more than 4% fly ash is impossible due to the swelling effect that occurs during binding.

Figure 7. Swelling of the cement mortar with the addition of 18% fly ash.

Table 5 presents the results of the leachability of the hazardous substances and heavy metals from the crushed cement mortars with the addition of 4% and 18% FA. The obtained results were compared with the highest permissible values for contaminants introduced to the water environment (Order of the Minister of Marine Economy and Inland Navigation 2019) [54]. The analyzed cement mortars had a highly alkaline reaction—above pH 11—which may result in high immobilization of heavy metals. The mobility of heavy metals is controlled by pH. In a highly alkaline environment (pH $\geq$ 11), FA could solidify/stabilize the majority of tested metals [55–57]. Out of all of the tested parameters (i.e., Cl^- SO_4^{2-}, NH_4^+, P, K, Ca, Li, Na), excessive values—about 70.47 mg/L—were noted only for sulfates (SO_4^{2-}) in samples with 18% fly ash. The leachability of heavy metals in the tested cement mortars was very low—below the limit of quantification. Only for barium did the values exceed the highest acceptable value (Order of the Minister of Marine Economy and Inland Navigation 2019) [54].

Table 5. The leachability of hazardous substances and heavy metals from cement mortars with the addition of fly ash, expressed in mg/L.

Parameter	Symbol	CEM I + 4% FA	CEM I + 18% FA	Highest Permissible Value [54,58]
pH	pH	11.6	11.1	6.0–9.0
Chloride	Cl^-	0.14	0.28	1000
Sulfate	SO_4^{2-}	455.28	570.47	500
Ammonium nitrogen	NH_4^+	blq **	blq **	10
Phosphorus	P	<0.005	<0.005	2
Potassium	K	5.83	8.75	80
Calcium	Ca	144.00	230.80	nr *
Lithium	Li	1.10	1.30	nr *
Sodium	Na	9.77	13.40	800
The sum of chloride and sulfate	$(Cl^- + SO_4^{2-})$	455.42	570.75	1500
Barium	Ba	15.20	35.50	2
Zinc	Zn	<0.10	<0.10	2
Copper	Cu	<0.20	<0.20	0.5
Lead	Pb	<0.50	<0.50	0.5
Cadmium	Cd	<0.50	<0.50	nr *
Chrome	Cr	<0.50	<0.50	0.1
Cobalt	Co	<0.05	<0.05	1
Iron	Fe	<0.04	<0.04	10
Manganese	Mn	<0.20	<0.20	nr *
Nickel	Ni	<0.40	<0.40	0.5

* No requirements, ** blq—values below the limit of quantification.

5.3. Evaluation of the Degree of Immobilization of Contaminants from Cement Mortars Modified with Nanomaterials with the Addition of Fly Ash

In many kinds of research, it is indicated [59,60] that the mechanical properties of cement mortars may be improved with the addition of nanosilica. At the next stage of this research, cement mortars with the addition of 4% and 8% FA were modified with nanosilica. Figures 8 and 9 present cement mortars with 4% and 8% fly ash from the incineration plant, respectively, modified with nanosilica.

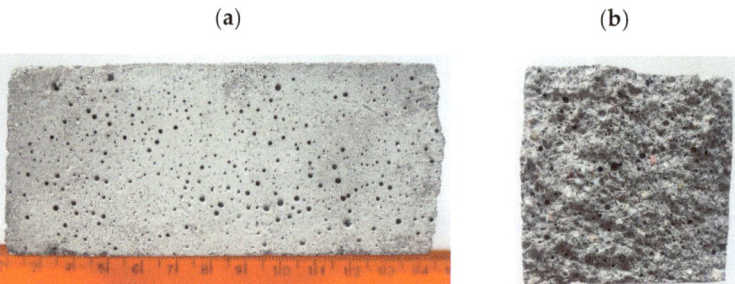

Figure 8. Cement mortar modified with the addition of 4% fly ash modified with nanosilica: (a) outer surface; (b) internal structural view.

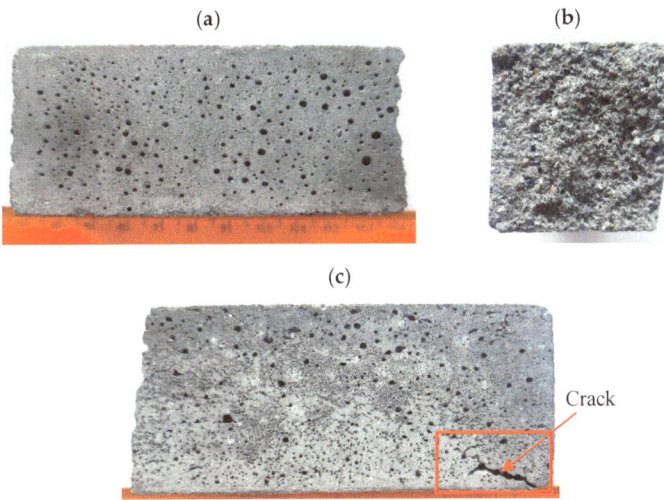

Figure 9. Cement mortar with the addition of 8% fly ash modified with nanosilica: (**a**) outer surface; (**b**) internal structural view; (**c**) crack in the cement mortar.

The results of the compressive strength and flexural strength tests of the concrete mortars with the addition of fly ash and modified with nanosilica after 28 days of curing are presented in Figure 10. In the case of mortars with the nanomaterial additive (CEM I + 4% FA + N), the strength of the samples was similar to or lower than the strength of samples without the nanomaterial added. The compressive strength of CEM I + 4% FA + N was 69.78 MPa, which is about 1.5 MPa less than that of the mortar without the nanomaterial additive. Its flexural strength was at the level of 9.00 MPa. In the case of mortars with 8% FA modified with nanosilica, the compressive strength and flexural strength were 35.06 MPa and 5.57 MPa, respectively. After modifying the mortar with nanosilica, its flexural strength increased by about 1 MPa. During visual analysis, cracks were noted in the CEM I + 8% FA + N mortar (Figure 9c), limiting its application in the construction industry.

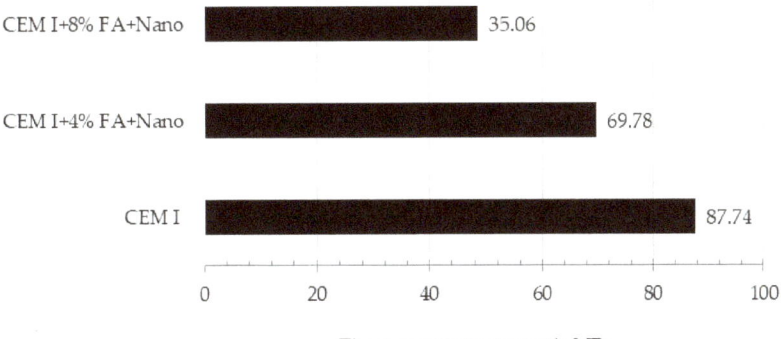

Figure 10. *Cont.*

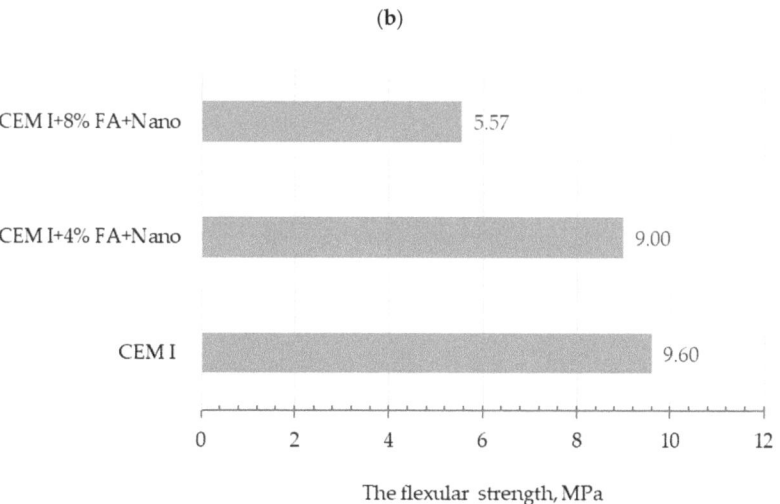

Figure 10. Strength of cement mortars with the addition of FA modified with nanosilica after 28 days of curing: (**a**) compressive strength; (**b**) flexural strength.

Table 6 presents the leachability of hazardous and heavy metals from cement mortars with the addition of FA modified with nanosilica. The discussed cement mortars had a strong alkaline reaction—above pH 11. It was noted that none of the tested parameters (i.e., Cl^-, SO_4^{2-}, NH_4^+, P, K, Ca, Li, Na) exceeded the permissible values determined in the Order of the Minister of Marine Economy and Inland Navigation 2019.

Table 6. Leachability of hazardous substances and heavy metals from cement mortars with the addition of fly ash modified with nanosilica, expressed in mg/L.

Parameter	Symbol	CEM I + 4% FA + Nano	CEM I + 8% FA + Nano	Highest Permissible Value [54,58]
pH	pH	11.8	11.1	6.0–9.0
Chloride	Cl^-	0.28	0.28	1000
Sulfate	SO_4^{2-}	285.24	331.86	500
Ammonium nitrogen	NH_4^+	blq **	blq **	10
Phosphorus	P	<0.005	<0.005	2
Potassium	K	4.01	6.70	80
Calcium	Ca	154.30	145.60	nr *
Lithium	Li	1.10	1.10	nr *
Sodium	Na	9.06	11.5	800
The sum of chloride and sulfate	($Cl^- + SO_4^{2-}$)	285.52	332.14	1500
Barium	Ba	8.70	17.80	2
Zinc	Zn	<0.10	<0.10	2
Copper	Cu	<0.20	<0.20	0.5
Lead	Pb	<0.50	<0.50	0.5
Cadmium	Cd	<0.05	<0.05	nr *
Chrome	Cr	<0.50	<0.50	0.1
Cobalt	Co	<0.50	<0.50	1
Iron	Fe	<0.04	<0.04	10
Manganese	Mn	<0.20	<0.20	nr *
Nickel	Ni	<0.40	<0.40	0.5

* No requirements, ** blq—values below the limit of quantification.

The contents of heavy metals were below the limit of quantification, except for the barium content, which was excessive. The barium content in the cement mortar with 4%

FA modified with the nanosilica was two times lower than that of the mortar without the addition of nanosilica.

6. Conclusions

Growing difficulties in the coal market, the uncertainty of supplies, and the dynamics of price changes in energy carriers give rise to legitimate concerns in the heating sector. Waste management regulations, including those for municipal waste, meet the needs of the market, providing space for the energetic use of the combustible fraction of municipal waste as fuel in combined heat and power plants and heating plants. However, such solutions translate into the formation of post-process waste, which is often classified as hazardous. One such byproduct is fly ash, whose chemical properties, toxicity, and dust emissions make it impossible to deposit in landfills for hazardous wastes. It should be noted that the actual method of neutralization of these wastes—i.e., depositing in the salt mine pits in Germany—slowly depletes. Looking to the future, it is necessary to search for alternative methods for the management of this waste that are neutral for the environment, economically justifiable, and compliant with the principles of the circular economy. The research conducted in this study is a natural response to the needs of the industry, which currently has limited possibilities for managing environmentally harmful waste products.

Based on this research, the following conclusions can be drawn:

- Tested fly ash from the MSWI plant was characterized by high leachability of sulfur ions, chlorides, calcium, sodium, zinc, lead, and chrome.
- The addition of fly ash in cement mortars increased the water demand. The use of fly ash as a partial replacement for cement will require the addition of more water or appropriate chemicals when designing the concrete mix.
- Replacing 4% of the cement mass with fly ash results in a slight reduction in the setting time of the cement grout (the beginning of the setting time guaranteed by the manufacturer is 186 min).
- Using nanosilica as an additive to cement mortar with fly ash does not improve its compressive and tensile strength.
- Test results proved the high immobilization of hazardous compounds by the C-S-H phase of the concrete. The leachability test confirmed the almost complete immobilization of chlorides and heavy metals by the C-S-H phase. The degree of immobilization exceeded 99%. Additionally, the leachability of the sulfate was limited to the level of 96–97%. The presented results are preliminary tests in a program designed to limit the impact of contaminants from the waste generated in the process of incineration of the mixed municipal waste fraction.
- The addition of nanosilica reduced the leaching of harmful substances (e.g., Cl^-, SO_4^{2-}, Ba) from cement mortar with 4% and 8% municipal waste fly ash. On the other hand, heavy metal immobilization was very high (99.9%). The level of immobilization did not depend on the addition of nanosilica to the mortar; in both analyzed variants, it was at a high level.
- Taking into account the current requirements for types of cement, fly ash from the incineration of municipal waste could be used only in small amounts (4%) for special cement that is chemically resistant and not commercially available.
- In the next steps of this research, the designed mortar with the addition of fly ash should be tested in various environmental exposure classes according to the standard PN-B-06265:2018-10 [61] to determine whether the leachability parameters change with the alteration of the structural behavior of concrete in various exposure classes.
- In future research, chemical ash degassing should be carried out to eliminate its influence on the swelling and cracking of cement mortars. Furthermore, the research should be continued by introducing it to a geopolymer.

The research conducted in this study is a natural response to the needs of the industry, which currently has limited possibilities for managing environmentally harmful waste products.

Author Contributions: Conceptualization, M.C. and B.Ł.-P.; methodology, M.C. and B.Ł.-P.; formal analysis, M.C. and B.Ł.-P.; writing—original draft preparation, M.C., M.K.-S. and B.Ł.-P.; writing—review and editing, M.C., M.K.-S. and B.Ł.-P. All authors have read and agreed to the published version of the manuscript.

Funding: This publication was funded by a subsidy allocated (08/030/BK_22/0101) for the year 2022 to the Department of Technology and Installations for Waste Management, Silesian University of Technology.

Data Availability Statement: Not applicable.

Acknowledgments: The research described in the article was partly carried out as part of the project "Silesian University of Technology as a Center of Modern Education based on research and innovation" POWR.03.05.00-00-Z098/17.

Conflicts of Interest: The authors declare no conflict of interest. The funders did not play any role in the design of the study plan; in collecting the resulting data, analyzing the results, or interpreting data; in formatting the content of the manuscript; or in the decision to publish the results of the study.

References

1. Directive 2008/98/EC of the European Parliament and of the Council of 19 November 2008 on Waste and Repealing Certain Directives. Available online: https://eur-lex.europa.eu/legal-content/PL/TXT/?uri=celex%3A32008L0098 (accessed on 5 September 2022).
2. Act of 14 December 2012 on Waste (Dz. U. z 2022 r. poz. 699, 1250,1726, 2127). Available online: https://isap.sejm.gov.pl/isap.nsf/DocDetails.xsp?id=WDU20220000699 (accessed on 5 September 2022).
3. Statistics Poland. Environment in 2021. Available online: https://stat.gov.pl/en/topics/environment-energy/environment/environment-2021,8,4.html (accessed on 5 September 2022).
4. Czop, M.; Łaźniewska-Piekarczyk, B.; Kajda-Szcześniak, M. Analysis of the Possibility of Using Slags from the Thermal Treatment of Municipal Waste as Potential Component of Cement—Case Study. *Materials* **2021**, *14*, 6491. [CrossRef] [PubMed]
5. Communication from the Commission to the European Parliament; The Council; The European Economic and Social Committee and the Committee of the Regions. *The Role of Waste-to-Energy in the Circular Economy*; COM/2017/034 Final. Available online: https://eur-lex.europa.eu/legal-content/en/TXT/?uri=CELEX%3A52017DC0034 (accessed on 5 September 2022).
6. Ferreira, C.; Ribeiro, A.; Ottosen, L. Possible applications for municipal solid waste fly ash. *J. Hazard. Mater.* **2003**, *96*, 201–216. [CrossRef] [PubMed]
7. Ghosh, S.K.; Kumar, V. *Circular Economy and Fly Ash Management*; Springer: Singapore, 2020. [CrossRef]
8. Rada, E.C.; Ragazzi, M.; Torretta, V.; Castagna, G.; Adami, L.; Cioca, L.I. Circular economy and waste to energy. *AIP Conf. Proc.* **2018**, *1968*, 030050. [CrossRef]
9. Sow, M.; Hot, J.; Tribout, C.; Cyr, M. Improving circular economy by the valorization of non-conventional coal fly ashes in composite cement manufacturing. *Constr. Build. Mater.* **2021**, *300*, 124053. [CrossRef]
10. Strzałkowska, E. Fly ash—A valuable material for the circular economy. *Gospod. Surowcami Miner. –Miner. Resour. Manag.* **2021**, *37*, 49–62. [CrossRef]
11. Czop, M.; Łaźniewska-Piekarczyk, B. Evaluation of the Leachability of Contaminations of Fly Ash and Bottom Ash from the Combustion of Solid Municipal Waste before and after Stabilization Process. *Sustainability* **2019**, *11*, 5384. [CrossRef]
12. Bertolini, L.; Carsana, M.; Cassago, D.; Curzio, A.Q.; Collepardi, M. MSWI ashes as mineral additions in concrete. *Cem. Concr. Res.* **2004**, *34*, 1899–1906. [CrossRef]
13. Ferraris, M.; Salvo, M.; Ventrella, A.; Buzzi, L.; Veglia, M. Use of vitrified MSWI bottom ashes for concrete production. *Waste Manag.* **2009**, *29*, 1041–1047. [CrossRef]
14. Forteza, R.; Far, M.; Segui, C.; Cerda, V. Characterization of bottom ash in municipal solid waste incinerators for its use in road base. *Waste Manag.* **2004**, *24*, 899–909. [CrossRef]
15. Ginés, O.; Chimenos, J.M.; Vizcarro, A.; Formosa, J.; Rosell, J.R. Combined use of MSWI bottom ash and fly ash as aggregate in concrete formulation: Enviromental and mechanical considerations. *J. Hazard. Mater.* **2009**, *169*, 643–650. [CrossRef]
16. Müller, U.; Rübner, K. The microstructure of concrete made with municipal waste incinerator bottom ash as an aggregate component. *Cem. Concr. Res.* **2006**, *36*, 1434–1443. [CrossRef]
17. Neville, A.M. *Properties of Concrete*; Longman: Harlow, UK, 1998.
18. Pera, J.; Coutaz, L.; Ambroise, J.; Chababbet, M. Use of incinerator bottom ash in concrete. *Cem. Concr. Res.* **1997**, *27*, 1–5. [CrossRef]
19. Sorlini, S.; Abba, A.; Collivignarelli, C. Recovery of MSWI and soil washing residues of concrete aggregates. *Waste Manag.* **2011**, *31*, 289–297. [CrossRef] [PubMed]
20. Ghazali, N.; Muthusamy, K.; Wan Ahmad, S. Utilization of Fly Ash in Construction. *IOP Conf. Ser. Mater. Sci. Eng.* **2019**, *601*, 012023. [CrossRef]

21. Jayaranjan, M.L.D.; van Hullebusch, E.D.; Annachhatre, A.P. Reuse options for coal fired power plant bottom ash and fly ash. *Rev. Environ. Sci. Biotechnol.* **2014**, *13*, 467–486. [CrossRef]
22. Marinina, O.; Nevskaya, M.; Jonek-Kowalska, I.; Wolniak, R.; Marinin, M. Recycling of Coal Fly Ash as an Example of an Efficient Circular Economy: A Stakeholder Approach. *Energies* **2021**, *14*, 3597. [CrossRef]
23. Rafieizonooz, M.; Khankhaje, E.; Rezania, S. Assessment of environmental and chemical properties of coal ashes including fly ash and bottom ash, and coal ash concrete. *J. Build. Eng.* **2022**, *49*, 104040. [CrossRef]
24. Rafieizonooz, M.; Mirza, J.; Salim, M.R.; Hussin, M.W.; Khankhaje, E. Investigation of coal bottom ash and fly ash in concrete as replacement for sand and cement. *Constr. Build. Mater.* **2016**, *116*, 15–24. [CrossRef]
25. Sow, M.; Hot, J.; Tribout, C.; Cyr, M. Characterization of Spreader Stoker Coal Fly Ashes (SSCFA) for their use in cement-based applications. *Fuel* **2015**, *162*, 224–233. [CrossRef]
26. Strzałkowska, E. The composition of the organic and inorganic matter of the siliceous fly ashes as part of their usefulness in technologies of building materials. *Gospod. Surowcami Miner. Miner. Resour. Manag.* **2016**, *32*, 71–88. [CrossRef]
27. Uliasz-Bocheńczyk, A.; Pawluk, A.; Sierka, J. Leaching of pollutants from fly ash from the combustion of biomass. *Gospod. Surowcami Miner. Miner. Resour. Manag.* **2015**, *31*, 145–1564.
28. Regulation of the Minister of Development of 21 January 2016 on the Requirements for the Thermal Treatment of Waste and the Methods of Handling Waste Generated as a Result of this Process (Dz.U. 2016, poz. 108). Available online: https://isap.sejm.gov.pl/isap.nsf/DocDetails.xsp?id=WDU20160000108 (accessed on 5 September 2022).
29. Bożym, M. The assessment of heavy metal binding forms in foundry wastes used as raw materials in agrotechnics, construction and road construction. *Gospod. Surowcami Miner. Miner. Resour. Manag.* **2022**, *38*, 169–189. [CrossRef]
30. Grünhäuser Soares, E.; Castro-Gomes, J.; Sitarz, M.; Zdeb, T.; Hager, I. The Immobilization of Heavy Metals from Sewage Sludge Ash in CO2-Cured Mortars. *Sustainability* **2021**, *13*, 12893. [CrossRef]
31. Giergiczny, Z.; Król, A. Immobilization of heavy metals (Pb, Cu, Cr, Zn, Cd, Mn) in the mineral additions containing concrete composites. *J. Hazard. Mater.* **2008**, *160*, 247–255. [CrossRef] [PubMed]
32. Kamal, N.L.M.; Beddu, S.; Syamsir, A.; Mohammad, D.; Itam, Z.; Hamid, Z.A.A.; Manan, T.S.A. Immobilization of Heavy Metals for Building Materials in the Construction Industry—An Overview. *Mater. Today Proc.* **2019**, *17*, 787–791. [CrossRef]
33. Król, A.; Jagoda, D. Carbonation and the strength properties of cement composites immobilizing heavy metals (Zn2+ Cr6+ Pb2+). *Cement Lime Concr.* **2012**, *17*, 90–101.
34. Haugsten, K.E.; Gustavson, B. Environmental properties of vitrified fly ash from hazardous and municipal waste incineration. *Waste Manag.* **2000**, *20*, 167–176. [CrossRef]
35. Polettini, A.; Pomi, R.; Sirini, P.; Testa, F. Properties of Portland cement—Stabilised MSWI fly ashes. *J. Hazard. Mater.* **2001**, *88*, 123–138. [CrossRef]
36. Sun, X.; Li, J.; Zhao, X.; Zhu, B.; Zhang, G. A review on the management of municipal solid waste fly ash in American. *Procedia Environ. Sci.* **2016**, *31*, 535–540. [CrossRef]
37. Regulation of the Minister of Climate of 2 January 2020 on the Waste Catalog (Dz.U. 2020, poz. 10). Available online: https://isap.sejm.gov.pl/isap.nsf/DocDetails.xsp?id=WDU20200000010 (accessed on 5 September 2022).
38. 2000/532/EC: Commission Decision of 3 May 2000 Replacing Decision 94/3/EC Establishing a List of Wastes Pursuant to Article 1(a) of Council Directive 75/442/EEC on Waste and Council Decision 94/904/EC Establishing a List of Hazardous Waste Pursuant to Article 1(4) of Council Directive 91/689/EEC on Hazardous Waste (Notified under Document Number C(2000) 1147) (Text with EEA Relevance). Available online: https://eur-lex.europa.eu/legal-content/EN/ALL/?uri=CELEX%3A32000D0532 (accessed on 5 September 2022).
39. PN-EN 15935:2022-01; Soil, Waste, Treated Bio-Waste and Sewage Sludge—Determination of Loss on Ignition (LOI). Polish Committee for Standardization: Warszawa, Poland, 2022.
40. PN-EN 450-1:2012; Fly Ash for Concrete-Part 1: Definitions, Specifications and Conformity Criteria. Polish Committee for Standardization: Warszawa, Poland, 2014.
41. Regulation of the Minister of Economy of 16 July 2015 on the Admission of Waste for Landfill Disposal (Dz.U. 2015 poz. 1277). Available online: https://isap.sejm.gov.pl/isap.nsf/DocDetails.xsp?id=WDU20150001277 (accessed on 5 September 2022).
42. PN-EN 12457-2:2006; Characterization of Waste-Leaching-Compliance Test for Leaching of Granular Waste Materials and Sludges. Part 2: One Stage Batch Test at a Liquid to Solid Ratio of 10 L/kg for Materials with Particle Size below 4 mm (without or with Size Reduction). Polish Committee for Standardization: Warszawa, Poland, 2006. Available online: https://sklep.pkn.pl/pn-en-12457-2-2006p.html (accessed on 15 April 2019).
43. PN-EN ISO 10523:2012; Water Quality—Determination of pH. Polish Committee for Standardization: Warszawa, Poland, 2012. Available online: http://sklep.pkn.pl/pn-en-iso-10523-2012e.html (accessed on 15 April 2019).
44. PN EN 27888: 1999; Water Quality-Determination of Electrical Conductivity. Polish Committee for Standardization: Warszawa, Poland, 1999.
45. PN-ISO 9297:1994; Determination of Chloride Ion Concentration by Titration (Mohr's Method). Polish Committee for Standardization: Warszawa, Poland, 1994. Available online: http://sklep.pkn.pl/pn-iso-9297-1994p.html (accessed on 15 April 2019).
46. PN-ISO 9280:2002; Determination of Sulphates (VI). Gravimetric Method with Barium Chloride. Polish Committee for Standardization: Warszawa, Poland, 2002. Available online: http://sklep.pkn.pl/pn-iso-9280-2002p.html (accessed on 15 April 2019).

47. *PN-ISO 9964-3:1994*; Determination of Sodium, Potassium, Calcium, Lithium and Bar by Flame Photometry. Polish Committee for Standardization: Warszawa, Poland, 1994. Available online: http://sklep.pkn.pl/pn-iso-9964-3-1994p.html (accessed on 15 April 2019).
48. *PN-EN ISO 6878:2006*; Determination of Phosphorus. Ammonium Molybdate Spectrometric Method. Polish Committee for Standardization: Warszawa, Poland, 2006. Available online: http://sklep.pkn.pl/pn-en-iso-6878-2006p.html (accessed on 15 April 2019).
49. *PN-EN 197-1:2012*; Cement—Part 1: Composition, Specifications and Conformity Criteria for Common Cements. Polish Committee for Standardization: Warszawa, Poland, 2012. Available online: https://sklep.pkn.pl/pn-en-197-1-2012p.html (accessed on 1 July 2021).
50. *PN-EN 196-1:2016-07*; Methods of Testing Cement-Part 1: Determination of Strength. Polish Committee for Standardization: Warszawa, Poland, 2018. Available online: http://sklep.pkn.pl/pn-en-196-1-2016-07e.html (accessed on 15 April 2019).
51. *PN-EN 196-3c: 2016-12*; Methods of Testing Cement-Part 3: Determination of Setting Times and Soundness. Polish Committee for Standardization: Warszawa, Poland, 2018. Available online: https://sklep.pkn.pl/pn-en-196-3-2016-12p.html (accessed on 1 July 2021).
52. *PN-EN 12457-4:2006*; Characterization of Waste-Leaching-Compliance Test for Leaching of Granular Waste Materials and Sludges. Part 4: One Stage Batch Test at a Liquid to Solid Ratio of 10 L/kg for Materials with Particle Size below 10 mm (without or with Size Reduction). Polish Committee for Standardization: Warszawa, Poland, 2006. Available online: https://sklep.pkn.pl/pn-en-12457-4-2006p.html (accessed on 15 April 2019).
53. Council Decision of 19 December 2002 Establishing Criteria and Procedures for the Acceptance of Waste at Landfills Pursuant to Article 16 of and Annex II to Directive 1999/31/EC. Available online: https://eur-lex.europa.eu/legal-content/GA/TXT/?uri=celex:32003D0033 (accessed on 5 September 2022).
54. Regulation of the Minister of Maritime Affairs and Inland Navigation of 12 July 2019 on Substances Particularly Harmful to the Aquatic Environment and on Conditions to Be Met When Discharging Wastewater into Waters or onto the Ground and When Discharging Rainwater or Snowmelt into Waters or into Water Installation (Dz.U. 2019 poz. 1311). Available online: https://isap.sejm.gov.pl/isap.nsf/DocDetails.xsp?id=WDU20190001311 (accessed on 5 September 2022).
55. Yakubu, Y.; Zhou, J.; Ping, D.; Shu, Z.; Chen, Y. Effects of pH dynamics on solidification/stabilization of municipal solid waste incineration fly ash. *J. Environ. Manag.* **2018**, *207*, 243–248. [CrossRef]
56. Pan, S.; Ding, J.; Peng, Y.; Lu, S.; Li, X. Investigation of Mechanochemically Treated Municipal Solid Waste Incineration Fly Ash as Replacement for Cement. *Energies* **2022**, *15*, 2013. [CrossRef]
57. Weibel, G.; Eggenberger, U.; Schlumberger, S.; Mäder, U.K. Chemical associations and mobilization of heavy metals in fly ash from municipal solid waste incineration. *Waste Manag.* **2017**, *62*, 147–159. [CrossRef] [PubMed]
58. Council Directive of 21 May 1991 Concerning Urban Waste Water Treatment. Available online: https://eur-lex.europa.eu/legal-content/PL/ALL/?uri=CELEX%3A31991L0271 (accessed on 5 September 2022).
59. Horszczaruk, E. Role of nanosilica in the formation of the properties of cement composites, state of the art. *Cement Lime Concrete* **2018**, *6*, 487–495.
60. Horszczaruk, E.; Mijowska, E.; Cendrowski, K.; Sikora, P. Influence of the new method of nanosilica addition on the mechanical properties of cement mortars. *Cement Lime Concrete* **2014**, *5*, 308–316.
61. *PN-B-06265:2018-10*; Concrete–Requirements, Properties, Production and Conformity–National Supplement PN-EN 206+A1:2016-12. Polish Committee for Standardization: Warszawa, Poland, 2016.

Article

Fly Ash as an Ingredient in the Contaminated Soil Stabilization Process

Kamil Banaszkiewicz *, Tadeusz Marcinkowski and Iwona Pasiecznik

Faculty of Environmental Engineering, Wroclaw University of Science and Technology, 27 Wybrzeże Wyspiańskiego St., 50-370 Wroclaw, Poland; tadeusz.marcinkowski@pwr.edu.pl (T.M.); iwona.pasiecznik@pwr.edu.pl (I.P.)
* Correspondence: kamil.banaszkiewicz@pwr.edu.pl

Abstract: Fly ash is the main by-product of coal combustion characterized by a large specific surface area. In addition to oxides, it also contains unburned coal and trace elements. This study aimed to investigate the possibility of using fly ash from pit-coal combustion (CFA) for the treatment of benzene-contaminated soil (S). The CFA was used as a mixture with Portland cement (PC) (70% PC + 30% CFA). The soil was treated with a PC-CFA mixture in amounts of 40, 60, and 80% of soil mass. During the process, the concentration of benzene was monitored with the flame-ionization detector. Produced monoliths (S+(PC-CFA)x) were tested for compressive strength and capillary water absorption. The experiment confirmed that the PC-CFA mixture limited benzene emission. The highest reduction in benzene concentration (34–39%) was observed for samples treated with the PC-CFA mixture in an amount of 80% (S+(PC-CFA)80). The average compressive strength of monoliths S+(PC-CFA)40, S+(PC-CFA)60, and S+(PC-CFA)80 was 0.57, 4.53, and 6.79 MPa, respectively. The water absorption values were in the range of 15–22% dm.

Keywords: hazardous waste; VOC; physicochemical stabilization; encapsulation; remediation

1. Introduction

Fly ash is a type of waste generated in coal power plants. Currently, about 35% of global electricity is produced from coal [1]. The largest coal-consuming economies are China (nearly half of global coal consumption), India, and the United States. Approximately 60% of electricity in China is produced from coal [2]. In addition, in Poland, electricity production is based on fossil fuels. In 2018, 48% of produced electricity came from hard coal, 29% from lignite, 13% from renewable sources, and 7% from gas [3,4]. Recently, a decreasing trend in worldwide coal consumption has been observed (Figure 1). In 2020, the decrease in coal consumption was 4% [5]. The main reasons for such changes are the industrial transformation and the COVID-19 pandemic. In 2020, the world energy demand decreased by 4.5%. In parallel, the CO_2 emissions from energy use fell by 6.3% [5]. It is estimated that in 2021, energy and coal consumption increased by 4.1% and 5% for G20 countries, respectively [6].

The combustion of coal results in large volumes of solid coal combustion products (CCPs). Shahzad Baig and Yousaf [7] reported that for every 4 tons of burnt coal, 1 ton of CCPs is produced. The total volume of CCPs generated globally in 2016 was approximately 1122 million tons [8,9]. From 2011 to 2016, the worldwide production of CCPs increased by 44.4% (Table 1).

The main constituent of CCPs is fly ash. Yao et al. [11] reported that coal fly ash accounts for 5 to 20% of the mass of coal burned. In 2016, in European (EU-15) power plants, about 40 million tons of CCPs was generated [12]. The fly ash accounted for 63.8 wt.% of CCPs (25.7 million tons) [12]. In 2016, 3.26 million tons of coal fly ash was produced in Poland. By 2018, its production had decreased to 2.43 million tons [13,14].

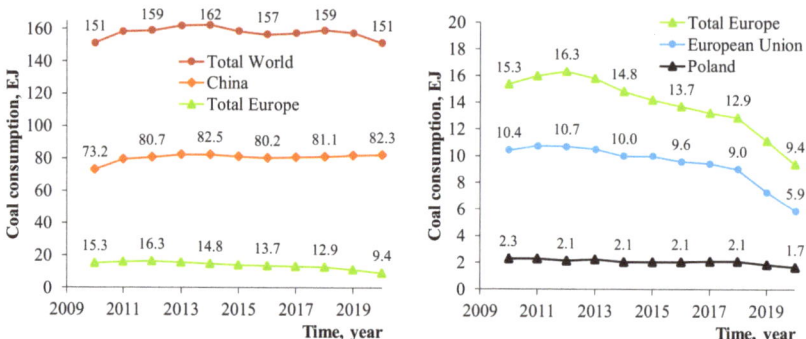

Figure 1. Coal consumption (data source comes from [5]).

Table 1. Production of CCPs by country/region [8–10].

Country/Region	CCPs Production, Mt		
	2011	2015	2016
Australia	13.1	12.2	12.3
Canada	6.8	6.2	4.8
China	395	565	565
Europe (EU)	52.6	105	140
India	105	240	197
Japan	11.1	12.6	12.3
Middle East and Africa	32.2	33.3	32.2
United States of America	118	117	107.4
Other Asia	16.7	25.3	29.6
Russian Federation	26.6	26.6	21.3
Total	777.1	1143.2	1121.9

In addition to CCPs, coal combustion contributes to the CO_2 emissions. The amount of CO_2 released depends on the type of coal burned [15]. In 2020, about 40% of CO_2 emissions (13.98 billion tons) came from coal [16].

The properties of fly ash are closely related to the type of coal burned and the combustion process. Fly ash with pozzolanic properties is produced during the combustion of hard, bituminous, and sub-bituminous coal. In the case of combustion of lignite with a higher lime and sulfur content, ash with hydraulic properties is also produced [9,17]. Fly ash is characterized by a large specific surface area, which ranges from 250 to 500 $m^2\ kg^{-1}$ [18–20]. It is mainly composed of SiO_2, Al_2O_3, Fe_2O_3, and CaO [21]. Fly ash also contains elements such as As, B, Cd, Co, Cr, Hg, Pb, Se, and organic constituents, e.g., PAHs and PCBs, which are responsible for its toxicity [18,22–24]. The inappropriate storage and utilization of fly ash may pose a threat to human health and the environment. Particularly during the landfilling of fly ash, toxic constituents can be released into the environment as dust or leachates [11,25–27]. The highest potential to contaminate waterways and soil is coal ash ponds. These ponds are a type of landfill whose bottom is unlined. As a result, the coal ash constituents can leach into the ground and water. The mobility of trace metals contained in fly ash mainly depends on pH. In turn, the pH of the water-fly ash system depends on the calcium content in fly ash. The alkalinity of fly ash may attenuate the mobility of Cd, Co, Cu, Hg, Ni, Pb, Sn, and Zn and enhance the mobility of As, B, Cr, Mo, Sb, Se, V, and W [28].

Due to the chemical composition of fly ash, its agriculture use is also limited. The use of large amounts of fly ash may adversely affect the soil's biological and chemical properties [29–31].

Fly ash is a by-product commonly used in the production of building materials. Moreover, due to its sorption properties, it can be used as a removal material of metals and some organic contaminants. A review of the coal fly ash's properties and use was presented by Ahmaruzzaman [32] and Yao et al. [11].

For example, the chemical composition of fly ash from coal combustion enables its use in clinker production. Moreover, its use as a raw material allows the reduction in the sintering temperature, thus saving energy [33,34]. It is an important application due to energy costs and CO_2 emissions. It is estimated that during the 1 kg of clinker production, 0.9–1 kg of CO_2 is emitted [35]. In 2020, cement production was responsible for 4.7% of global CO_2 emissions (1.63 billion tons) [16]. At that time, the global production of cement reached the level of 4.1 billion tons [36].

In the case of using fly ash in the production of cement, concrete, or low-cement binders, the pozzolanic properties are of particular importance. In Europe, the requirements for fly ash use in cement and concrete were described in the EN 450-1:2012 and EN 197-1:2011 [37,38] standard. The use of fly ash in concrete has a beneficial effect on its functional properties [32,39]. The addition of fly ash improves the workability of concrete. This is due to the spherical shape of its particles, which provides a lubricating effect in the concrete. The application of fly ash also improves the homogeneity of the concrete [17]. Moreover, fly ash, when used as a pozzolanic admixture, reacts with calcium hydroxide and forms a product similar in composition and properties to calcium silicate hydrate (C-S-H) [40,41]. In parallel, it follows a reduction in the pore interconnectivity of concrete, thus decreasing its permeability. The pozzolanic reactions are long-term processes; therefore, concrete containing fly ash may show a lower strength at a curing period of 28 days [39,42]. To improve the early-age strength, the development of fly ash concrete is possible by using an elevated curing temperature [43,44]. In addition to delaying binding and slowing down concrete hardening, a lower temperature rise due to less heat of hydration is observed [39,40]. The data on the fly ash pozzolanic properties and their reactions in concrete are presented in the ACI Committee report [39].

Another area of engineering application of fly ash is the production of cellular concrete [45–47], lightweight aggregates [48,49], ceramic tiles, and bricks [50]. The important sector that uses fly ash is mining. Fly ash is used for the back-filling of underground mining voids [51]. It is a typical method of fly ash recovery in Poland [52].

In 2016, in the EU-15 Member States, 31.5 wt.% of the total fly ash generated was used as a cement raw material (7.0%), as a constituent in blended cement (7.2%), and as an addition for the production of concrete (17.3%) [12]. Fly ash was also used in concrete blocks, reclamation, and for the infill of voids, mine shafts, and subsurface mine working [12,53]. In 2016, the EU-15 achieved a 94% utilization rate of fly ash [12]. In Polish power plants using hard coal, silicate fly ash is usually obtained. This material is a valuable mineral resource for the building materials industry. In 2018, Poland achieved a 90.4% recovery rate of coal fly ash [14].

An interesting application of fly ash is its use as an adsorbent. Due to the specific properties of fly ash (large specific surface area and porosity), they can be used as a sorbent for selected organic compounds, e.g., VOCs. The European Union Directive 2004/42/CE defines VOCs as any organic compound having an initial boiling point less than or equal to 250 °C measured at a standard pressure of 101.3 kPa [54]. The United States of America established a definition of VOCs based on photochemical reactivity. In accordance with the Code of Federal Regulations (40 CFR 51.100.), VOCs means any compound of carbon, excluding carbon monoxide, carbon dioxide, carbonic acid, metallic carbides or carbonates, and ammonium carbonate, which participates in atmospheric photochemical reaction [55]. The volatile organic compounds impose serious environmental problems such as global warming and ozone depletion [56]. One of the important classes of VOCs is aromatic hydrocarbons such as benzene, toluene, ethylbenzene, and xylenes (BTEX). Due to toxic, mutagenic, or carcinogenic (benzene, ethylbenzene) properties, the BTEX compounds are recognized as hazardous environmental pollutants [57]. The International Agency for Research on Cancer (IARC) classified benzene as carcinogenic to humans [58]. Benzene is used in the manufacture of styrene, phenols, cyclohexane, alkylbenzenes, and chlorobenzene. These substances may be used for the production of rubber, resins, nylon, or detergents. Benzene is also a natural constituent of gasoline. According to the Directive 2009/30/EC, its maximum content in petrol is limited to 1% v/v [59,60].

The BTEX compounds are emitted into the environment mainly from the chemical and petrochemical industry, motor vehicle transport, coal combustion, or painting [61–65]. BTEX compounds may diffuse into air and water. Furthermore, they penetrate the soil and cause a loss in soil functionality [66]. They can be adsorbed by organic matter occurring in soil, and also onto soil mineral surfaces.

One of the methods that can be used for the treatment of soil containing inorganic and organic (e.g., BTEX) contaminants is the technology of chemical solidification. In this process, the waste is blended with cement, fly ash, or their mixtures. This technique is widely used for the treatment of waste contaminated with heavy metals [67]. The inorganic contaminants show compatibility with cement [68]. As a result, the reduction in water content, increase in strength, and lowered mobility of contaminants are observed. Moreover, this method is used, e.g., in construction, to stabilize soils [69]. In the case of treating waste contaminated with organic compounds, the usefulness of this technique may be limited [70]. The organic compounds do not react with cement and fly ash. However, they can affect the efficiency of the cement and fly ash hydration process [68,71]. As a result, the products may show lower compressive strength [72,73].

Despite the application of cement and fly ash for utilization waste contaminated with organic, there is a lack of studies about the treatment of BTEX-contaminated soils, especially concerning their amount and characteristics of emissions during the treatment process. In the chemical solidification process, the BTEX contaminants can be physically trapped in a hardened binder or adsorbed by additional binding mixture constituents, e.g., activated carbon. For example, Butler et al. [74] demonstrated that the physical encapsulation of toluene is possible. The entrapped toluene formed vesicular, randomly placed structures in cement. To improve the BTEX encapsulation process, it is necessary to use the medium that will adsorb them and reduce their migration. The additives used for this purpose are fly ash, organoclays, or activated carbon.

The objectives of this study were (1) to investigate the possibility of using fly ash as a low-cost material for the treatment of benzene-contaminated soil (the fly ash was used as a mixture with cement); (2) the analysis of the variations in the benzene level emitted during the process; (3) the evaluation of physical properties of produced monoliths.

2. Materials and Methods

2.1. Reagents for Preparation of Stabilizing Mixtures

The following materials were used in the benzene-contaminated soil stabilization research described in this paper: Portland cement CEM I 45.5 R and fly ash from coal-burning. The cement used in the experiment was characterized by strength class 42.5 and high early strength (CEM I 42.5 R). The material met the requirements of the EN 197-1 standard [38]. A characteristic of PC is the production of a significant amount of heat during the setting. As a result, it can be used at low ambient temperatures.

The second component of the prepared stabilizing mixture was fly ash obtained from a Lower Silesian coal-fired power plant. The CFA contained around 53% SiO_2, 24% Al_2O_3, 6.4% Fe_2O_3, 3.4% CaO, 2.8% MgO, and 0.5% SO_3. The important chemical parameter of fly ash is loss on ignition (LOI). It has to be noted that LOI does not represent the unburned carbon in ashes, because of the presence of several other compounds that also decompose on heating. The high content of unburned carbon increases the water demand of the fly ash. As a result, the products based on fly ash with high unburned coal content show low strength. The loss on ignition of used CFA was 3.9%.

2.2. Soil

The research was carried out on soil from rural areas in Lower Silesia (Poland). The basic material properties, such as humidity and organic matter content, were 3.41% (by weight) and 1.97% of dry matter, respectively. The particle-size distribution analysis showed that the soil was noncohesive (Figure 2). The sand fraction content was greater than 90% (by weight). Sieve analysis also showed the presence of dust particles with grains below 0.063 mm in diameter (ca. 5.3%). The bulk density of soil amounted to 1.21 g mL^{-1}. The estimated particle diameters D10, D30, and D60 were, respectively, 0.12 mm, 0.30 mm, and 0.57 mm. Based on substitutive diameters, the uniformity coefficient (C_U = 4.91) (1) and coefficient of gradation (C_C = 1.34) (2) were calculated.

$$C_U = D60/D10, \tag{1}$$

$$C_C = D30^2/(D10 \cdot D60), \tag{2}$$

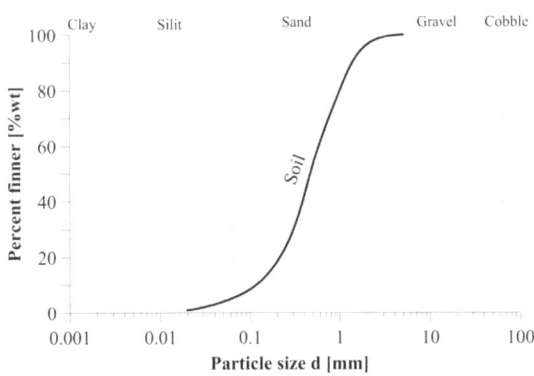

Figure 2. Grain size distribution curve of used soil.

According to the applied procedure, 200 g samples of raw soil were contaminated with 0.5 mL of benzene and mixed for 2 min using a Heidolph Reax 20/8 shaker. Soil samples prepared in this way were stored at 7 °C for 24 h and then treated with the PC-CFA mixture.

2.3. Methodology of Contaminated Soil Treatment Process

The benzene-spiked soil samples (S) were processed with the PC-CFA mixture. The content of PC and CFA in the mixture was, respectively, 70% w/w and 30%. In turn, the dose of the PC-CFA mixture was 40%, 60%, and 80% of the soil mass.

The homogenization of all components (S+PC-CFA) was performed for 5 min in a Tecnotest B205/X5 mixer. Monoliths with soil were obtained by the further addition of water for initiation of the cement hydration process. Homogenization of the wet mix also took 5 min. The obtained cement-soil mixture was molded and hardened for 28 days.

Temperature conditions have a significant influence on the emission of VOCs. For this reason, the temperature and relative humidity were monitored throughout the process. The values were, respectively, 20 ± 0.5 °C and 25 ± 3% RH. Measurements were made using the Hanna Instruments HI 9564 Thermohigrometer.

During the process, the concentration of benzene in the reactor was monitored. The amount of benzene was measured using a Micro FID flame-ionization detector (Photovac). The gas samples were transported to the detector through PTFE tubing. Moreover, a dust filter was installed at the inlet to the PTFE tubing. The FID was calibrated against methane. The benzene concentration was calculated based on reaction coefficients that may be applied in the situation of a single contaminant in the atmosphere. To evaluate the repeatability and stability of the process, all samples were tested three times.

The effectiveness of using the PC-CFA mixture as a stabilizing material for benzene-contaminated soil was assessed based on a comparative analysis of concentrations in the phase of homogenization of dry components and reference samples (S+0 (x)). In the case of zero samples, the PC-CFA mixture was not added. Their composition included only 200 g of raw soil contaminated with 0.5 mL of benzene. The parameters of homogenization of the reference samples were the same as those of the samples with the PC-CFA mixture.

2.4. Compressive Strength

One of the parameters tested to determine the effectiveness of the chemical solidification process was compressive strength. The prepared soil-cement-ash mixtures were placed in cylindrical molds and compacted. After 3 days of maturation, samples were de-molded and cured for another 25 days. The strength tests of the monoliths with benzene-contaminated soil (with diameter and height = 50 ± 3 mm) were determined after 28 days of hardening. Each of the prepared soil-cement-ash mixtures was tested three times.

2.5. Capillary Rise Height of Water

Capillary rise height was evaluated during the first phase of determining mass absorbability. The experiment was carried out to illustrate the behavior of soil-cement-ash monoliths deposited in landfills (this especially concerns bottom parts of monoliths exposed to long-term contact with eluates). To determine the speed of water imbibition, dried soil composites were submerged in water to $\frac{1}{4}$ of their height. During the test, water was absorbed mainly through cylindrical surfaces of monoliths. The height of water imbibition was measured after 15, 30, and 60 min.

2.6. Water Absorption

Mass water absorption is the ratio of the mass of water absorbed through the monolith to the weight of the dry monolith. The water absorption capacity test of monoliths was carried out by the procedure described in the PN-B-04101 standard [75]. According to the test method, samples were stepwise-immersed in water. Such a treatment is to avoid the entrapment of air inside the monolith. In the first step, dried monoliths were placed in a vessel and flooded with water to $\frac{1}{4}$ of their height. After 2 h, the water level was increased to $\frac{1}{2}$ of their height, and after another 3 h, up to $\frac{3}{4}$. The samples were left in such an immersion for 19 h, after which they were completely flooded with water. Finally, the upper surface of the samples was 2 cm below the water level. Such prepared samples were left for 24 h. Next, samples were taken from the water and weighed (with the accuracy of ±0.01 g) in 24 h intervals. The weighing was repeated until a constant weight of the monolith was achieved.

Water absorption (W_A) was calculated according to Equation (3):

$$W_A = [(W_w - W_d)/W_d] \cdot 100\%, \tag{3}$$

where: W_w is the mass of the monolith saturated with water and W_d is the mass of a monolith in the dry state.

3. Results and Discussion

3.1. Efficacy of the Cement-Fly Ash Matrix

Data on the benzene concentration changes in the mixer during the process are presented graphically in Figure 3.

The FID measurements showed a sudden increase in benzene concentration at the beginning of the mechanical mixing of components. Concentrations of benzene recorded by the FID were in the following ranges: (i) <0.1–2596 ppm C_6H_6 for zero samples, (ii) 1.8–2280 ppm C_6H_6 for series of samples stabilized with PC-CFA mixture in an amount of 40% of soil mass, (iii) <0.1–1923 ppm C_6H_6 for series of samples stabilized with PC-CFA mixture in an amount of 60% of soil mass, and (iv) 0.3–1486 ppm C_6H_6 for series of samples stabilized with PC-CFA mixture in an amount of 80% of soil mass. The highest concentrations were measured for reference samples (S+0 (x) samples). For all samples, the maximum concentration of benzene was recorded between the 30th and the 45th seconds of the process (Figure 3). The research confirmed that the risk of releasing significant amounts of gaseous contaminants occurs mainly during the homogenization of dry components (soil and PC-CFA mixture) [76,77].

The introduction of the PC-CFA mixture caused a decrease in maximum momentary benzene concentration. The level of the decrease was in the range of (i) 12–19% for S+(PC-CFA)40 samples, (ii) 24–35% for S+(PC-CFA)60 samples, and (iii) 40–46% for S+(PC-CFA)80 samples (Table 2). Nevertheless, the introduction of fine-grain-size materials (PC and CFA) did not influence the dynamics of the analyzed solvent release. In each trial, the emission of 90% of the total quantity of benzene released during the process occurred between 100 and 150 s of the homogenization phase. The properties of the solvent and used removal material have a significant influence on gaseous contaminant emission characteristics [77]. The main mechanisms of the discussed benzene removal technology are adsorption and physical encapsulation processes. Benzene is a nonpolar compound, which is slightly soluble in water [78]. In the soil environment, it can be adsorbed on soil particles or fill soil pores as a gas [79]. One of the factors determining its adsorption in the liquid–solid interface is soil organic matter content [80]. Lake et al. [81] reported that benzene interacts with hydrophobic groups of humic acid. The used soil was characterized by low organic matter content, which could affect the adsorption of benzene. Another factor that influences the adsorption efficacy is soil particle size. Sun et al. [82] showed that as the particle size decreases, the adsorption efficiency of benzene increases. It is a result of the increase in the specific surface area of the soil particles. In comparison to silty soils or clay soils, sandy soils characterize the lower specific surface area. The sand fraction content in the used soil was greater than 90% (by weight). Another factor that affected benzene release is the speed of mixing soil with stabilizing-binding components. The applied rotary speed was 140 rpm. During the mixing, the particles of contaminated soil violently collide. As a result, the increase in the processes of desorption and volatilization of vapors collected in soil pores is observed [82]. Thus, when treating benzene-spiked soil, lowering the mixing speed of dry components may have a positive impact on the initial volume of gaseous pollutants released.

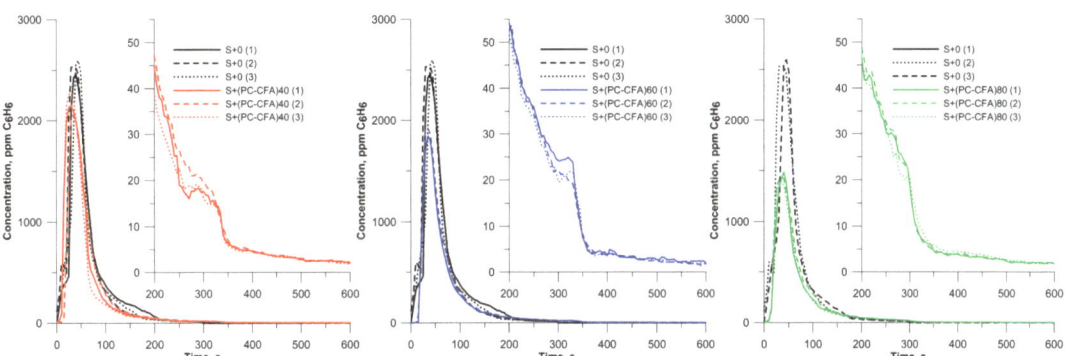

Figure 3. Comparison of benzene emissions during the chemical solidification process of soil using cement-fly ash mixture. Code of the sample: S—soil, PC-CFA—Portland cement-coal fly ash mixture, 40/60/80—PC-CFA mixture dose expressed as a percent of soil mass, (1)/(2)/(3)—sample number.

An important moment of the chemical solidification process is the hydration stage. The required amount of water depends on the quantity of cement and immobilizing additives, their properties, as well as the water demand of the processed waste. When treating soil contaminated with VOCs, limiting their release during the homogenization stage may result in a higher concentration of pollutants in the process air during the hydration stage. Donaldson et al. [83] reported that the introduction of water accelerates the removal from the soil of hydrocarbons found in gasoline. This phenomenon was observed in research of the neutralization of soils contaminated with ethylbenzene or xylenes [76,77]. Water fills the pores of the soil and blocks minerals' specific surface area. As a result, a lowering in the adsorption capacity of the organic vapors is observed [84]. In the research, water was added in the 300th second of the dry components' mixing. The temporary stabilization or a slight increase in benzene concentration was observed for samples treated with the PC-CFA mixture in amounts of 40 and 60% of soil mass (S+(PC-CFA)40 and S+(PC-CFA)60, respectively) (Figure 3).

Table 2. PC-CFA mixture efficacy.

Series	Sample	Maximum Momentary Concentration, ppm C_6H_6	Average Concentration, ppm C_6H_6			Immobilization Efficacy, %
			Homogenization Stage	Hydration Stage	Process	Homogenization Stage
(1)	S+0	2463	430	-	-	-
	S+(PC-CFA)40	2154	386	5.20	194	10
	S+(PC-CFA)60	1844	309	5.72	156	28
	S+(PC-CFA)80	1486	262	3.93	132	39
(2)	S+0	2531	450	-	-	-
	S+(PC-CFA)40	2060	363	5.53	183	19
	S+(PC-CFA)60	1923	337	5.61	170	25
	S+(PC-CFA)80	1438	289	3.98	145	36
(3)	S+0	2596	397	-	-	-
	S+(PC-CFA)40	2280	364	4.99	183	8.3
	S+(PC-CFA)60	1675	308	6.23	156	22
	S+(PC-CFA)80	1395	262	4.58	132	34

The continuous measurements showed that the most neuralgic stage of the benzene-contaminated soil chemical solidification process is the homogenization phase. The average concentration, as well as the total amount of emitted benzene in the first stage of the process, was several dozen times higher than in the hydration stage (Table 2). Nevertheless, the introduction of the PC-CFA mixture lowered the averaged benzene concentration in the reactor. A decrease in the concentration increased the PC-CFA dose.

The highest reduction in average benzene concentration in the homogenization stage (in the range 34–39%) was recorded for samples S+(PC-CFA)80. In the case of samples S+(PC-CFA)60, the benzene immobilization efficacy was in the range of 22–28%. The lowest effectiveness (from 8 to 19%) was observed for composites S+(PC-CFA)40 (Table 2). The main factor determining the effectiveness of the analyzed process is the unburned carbon content (UBC). The UBC mainly depends on fly ash dose. In turn, the carbon content in fly ash depends on the type of coal and its combustion conditions [85,86] Thus, the presented results are adequate only for similar process parameters (temperature, humidity, and soil and fly ash properties). An important element of the used treatment process is the interaction of cement with fly ash. During the hydration process, cement paste may physically block the internal fly ash pores [81]. As a result, its adsorption capacity may be limited. One of the factors that may improve the benzene adsorption may be the introduction of fly ash before cement addition. As a result, benzene associated with fly ash can be coated with hydrated cement [81]. In this way, its volatilization can be reduced.

The greatest problem in monitoring benzene emission during the chemical solidification process was high concentrations of particulate pollutants in the air inside the reactor. This was especially true for the stage of homogenizing binding components with soil. In the case of continuous measurements, mineral particles were eliminated in a filter mounted before a combustion chamber. Nevertheless, the limitation of process airflow by solid particles accumulated on the filter and in ducts might affect the quality of continuous measurements in the phase of hydration. In the conference paper [87], the authors compared the results of continuous measurements of benzene concentration in the reactor and its mass adsorbed on activated carbon.

3.2. Compressive Strength

Compressive strength is one of the parameters determining the efficiency of the chemical solidification process. To be suitable for deposition and transport, monoliths must be characterized by some minimum value of stress. Knowledge of the stress that does not change monoliths' shape and structure enables, among others, the determination of the maximum thickness of waste in the landfill. Unfortunately, no normative regulations according to the compressive strength of solidified waste are currently in force in Poland.

The compressive strength tests of solidified benzene-contaminated soil were performed after 28 days of hardening. Measurements were taken in three replications. The impact of the applied PC-CFA mixture dose on the strength of the monoliths with benzene-contaminated soil is illustrated in Figure 4.

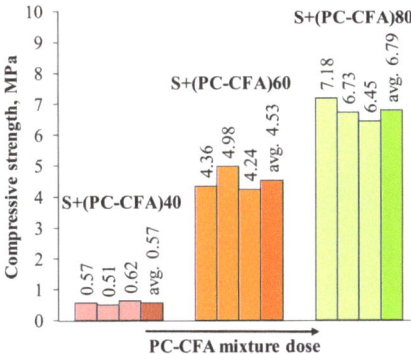

Figure 4. Dependence of the compressive strength of the monoliths with benzene-contaminated soil on PC-CFA mixture dose.

The average compressive strength (f_c) of monoliths S+(PC-CFA)40, S+(PC-CFA)60, and S+(PC-CFA)80 was 0.57 MPa, 4.53 MPa, and 6.79 MPa, respectively (with water–binder ratio 0.77, 0.64, and 0.54, respectively). The worst mechanical parameters (avg. f_c = 0.57 MPa) were found in samples solidified with the binding mixture in an amount of 40% of soil mass. Such a level of compressive strength corresponds to the characteristics of poor cement-lime mortar. Increasing the stabilizing-binding mixture dose from 40% to 80% of soil mass resulted in an almost twelvefold increase in strength value. The use of cement-fly ash mixtures is beneficial from a cost perspective and organic contaminants sorption. Nevertheless, it is known that cement-fly ash mixtures show a lower early compressive strength compared to the samples based on cement without fly ash. The fly ash reacts with cement hydrates (portlandite) and forms extra C-S-H gel that helps to achieve a more compact structure. Nevertheless, the rate of pozzolanic reaction in 28 days is slow. At a curing period of 28 days, the cement-fly ash mixtures are characterized by a higher percentage of voids. The strength is particularly influenced by the presence of voids bigger than 20 nm [88]. As a result, the strength of monoliths mainly depends on the amount of introduced cement. In addition, the presence of mineral particles smaller than 0.063 mm may be particularly important. This fraction increases water absorbability and sticks agglomerate grains, thus diminishing the zone of contact with pure agglomerate. The impact of introducing soil spiked with benzene should also be considered. Ezeldin et al. [89] showed that replacing fine aggregate in concrete with benzene-contaminated soils causes a decrease in mechanical properties of concrete, especially when silty sand was used. However, the pozzolanic reaction improves to a limited extent the strength of monoliths based on cement blended with fly ash with an increase in the curing time [90,91].

3.3. Capillary Rise Height of Water

In addition to a high degree of encapsulation of processed soil and sufficiently high compressive strength, monoliths must be characterized by low permeability and low liquid penetration. Both high mass absorbability and capillary pull-up have a destructive effect on the durability of products of the chemical solidification process. The properties mentioned are especially essential in the case of depositing solidified waste in landfills (this especially concerns bottom parts of waste exposed to long-term contact with eluates). In addition, the low tightness of monoliths may cause the secondary emission of organic pollutants' built-in matrices of hydraulic binders. Thus, the formation of monoliths characterized by high tightness is essential.

Soil-cement-ash monoliths elaborated in the experiment were characterized by the mass ratio of water to PC-CFA mixture from 0.54 for sample S+(PC-CFA)80 to 0.77 for sample S+(PC-CFA)40. The height of water imbibition was measured after 15, 30, and 60 min of contact with water. Analysis of capillary action after 28 days of maturation showed that all monoliths were characterized by the high content of capillary pores. In the case of sample S+(PC-CFA)40, water reached 3.5 cm high in only 60 min (Figure 5). Increasing the dose of the PC-CFA mixture to 80% of soil mass (S+(PC-CFA)80) resulted in a reduction in capillary rise height to 1.3 cm (after 60 min of contact with water). Long-term moistness of the solidified waste with liquids of high salinity may lead to the structural destruction of cement slurry. The high permeability of prepared monoliths might be a result of the formation of a higher percentage of voids in mixtures containing fly ash (after 28 days of maturation) and a poor bonding of benzene-contaminated soil with hydrated cement paste. Generally, the pore structure of cement paste is closely related to the water–cement ratio and hydration period [92].

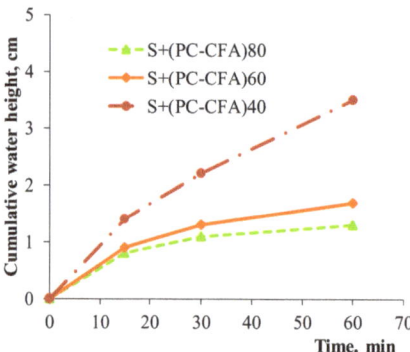

Figure 5. Cumulative height of water absorbed by capillarity of the solidified samples.

3.4. Water Absorption

Data on the water absorption capacity of monoliths with benzene-contaminated soil are shown in Figure 6. The test showed a decrease in the water absorption capacity of the samples with an increase in the proportion of the PC-CFA mixture. All the tested specimens based on the PC-CFA mixture exhibited high water absorption capacities by 15–22% dm. Samples solidified with the mixture in amounts of 60% and 80% of soil mass (S+(PC-CFA)60 and S+(PC-CFA)80, respectively) were characterized by similar values of water absorption, at the level of about 15–16% dm (Figure 6).

The highest value was presented by sample S+(PC-CFA)40, which was 22% dm. Moreover, the monolith S+(PC-CFA)40 was unstable upon water immersion. Degradation of the sample S+(PC-CFA)40 was observed on the 6th day of its immersion in water. High water absorption creates a significant risk in the case of exposure to sub-zero temperatures. An increased volume of water in pores during a phase transition weakens the structure of monoliths mainly through internal crack growth [93].

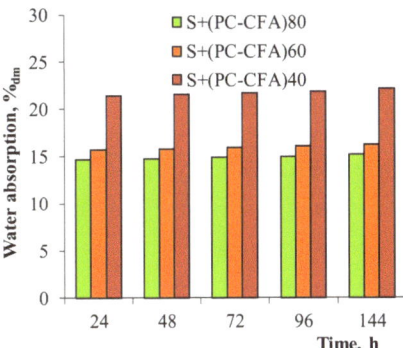

Figure 6. Water absorption capacity of solidified specimens.

4. Conclusions

The evaluated data led to the following conclusions:

- Continuous measurements of benzene concentrations showed that its emission had a very dynamic course. In all tests, the emission of 98% of the total amount of benzene released in the chemical solidification process already occurred in the homogenization phase.
- Experiments confirmed the possibility of limited use of the PC-CFA mixture as a low-cost material to remove benzene. The results showed that the PC-CFA mixture could limit the volume of benzene released during the process. The introduction of the cement-ash mixture in an amount of 80% of contaminated soil (by weight) reduced benzene emission in the first phase of the process by avg. 36%.
- The research did not show any significant influence of the amount of the stabilizing-binding mixture on emission dynamics of the analyzed solvent.
- The adsorption properties of the fly ash mainly depend on unburned carbon content (UBC). The UBC depends on the type of coal and its combustion conditions. Thus, the presented results are adequate only for the used fly ash type and process conditions (the proven efficiency of the PC-CFA mix applies to the following conditions: humidity of 25% RH and temperature of 20 °C).
- Mechanical tests confirmed a strict relationship between the value of mechanical compressive strength and the dose of binding materials. The highest strength (more than 6 MPa) was observed in samples that were treated with the PC-CFA mixture in an amount of 80% of the mass of processed soil. The possible explanation of low compressive strength values may be a poor bonding of benzene-contaminated soil with hydrated cement paste or slow pozzolanic reaction of fly ash.
- The capillary imbibition of water after 28 days of maturation showed that all monoliths were characterized by the high content of capillary pores.
- The water sorption capacity of the spiked soil sample solidified with the PC-CFA mixture in an amount of 40% of soil mass was 22% dm. The monolith S+(PC-CFA)40 was unstable, and its degradation was observed on the 6th day of its immersion in water.
- The increase in the dose of the ash-cement mixture resulted in a reduction in capillary imbibition of water and its absorption by immersion.

In conclusion, the cement-ash mixture can be considered a cheap and effective material for the treatment of soils containing low levels of benzene. The analyzed process can be used in developing countries without infrastructure for hazardous waste treatment. Nevertheless, due to the parameters that affect the effectiveness of benzene entrapping (e.g., temperature conditions, and soil and fly ash properties), the use of the PC-CFA mixture may be limited.

Author Contributions: Conceptualization, T.M. and K.B.; methodology, K.B. and T.M.; investigation, K.B. and I.P.; data analysis, K.B. and I.P.; writing—original draft preparation, K.B., I.P. and T.M.; writing—review and editing, K.B., I.P. and T.M. All authors have read and agreed to the published version of the manuscript.

Funding: This research received no external funding.

Institutional Review Board Statement: Not applicable.

Informed Consent Statement: Not applicable.

Data Availability Statement: Not applicable.

Conflicts of Interest: The authors declare no conflict of interest.

References

1. International Energy Agency. *Coal 2020. Analysis and Forecast to 2015*; IEA: Paris, France, 2020. Available online: https://www.iea.org/reports/coal-2020 (accessed on 7 January 2022).
2. International Energy Agency. China Has a Clear Pathway to Build a More Sustainable, Secure and Inclusive Energy Future–News–IEA. Available online: https://www.iea.org/news/china-has-a-clear-pathway-to-build-a-more-sustainable-secure-and-inclusive-energy-future (accessed on 7 January 2022).
3. Enerdata. Global Energy Statistical Yearbook 2019. 2019. Available online: https://yearbook.enerdata.net/coal-lignite/coal-world-consumption-data.html (accessed on 9 July 2020).
4. Ministerstwo Aktywów Państwowych. Krajowy Plan na Rzecz Energii i Klimatu na Lata 2021–2030 (Ministry of State Assets National Energy and Climate Plan for the Years 2021–2030). 2019. Available online: https://www.gov.pl/web/klimat/krajowy-plan-na-rzecz-energii-i-klimatu (accessed on 9 July 2020).
5. BP. Statistical Review of World Energy. 2021. Available online: https://www.bp.com/en/global/corporate/energy-economics/statistical-review-of-world-energy.html (accessed on 7 January 2022).
6. Enerdata. Global Energy Trends–2021 Edition. 2021. Available online: https://www.enerdata.net/publications/reports-presentations/world-energy-trends.html (accessed on 7 January 2022).
7. Shahzad Baig, K.; Yousaf, M. Coal Fired Power Plants: Emission Problems and Controlling Techniques. *J. Earth Sci. Clim. Chang.* **2017**, *8*, 404. [CrossRef]
8. World Wide Coal Combustion Network. Member Information 2018/2019 via: Harris et al. 2020. Available online: https://wwccpn.com/ (accessed on 7 January 2022).
9. Harris, D.; Heidrich, C.; Feuerborn, J. Global Aspects on Coal Combustion Products. *VGB PowerTech* **2020**, *10*, 25–33. Available online: https://www.vgb.org/vgbmultimedia/PT202010HARRIS-p-16422.pdf (accessed on 3 April 2021).
10. Heidrich, C.; Feuerborn, H.J. Coal Combustion Products: Global Operating Environment, WWCCPN, World of Coal Ash. 2017. Available online: https://www.adaa.asn.au/blog/2017/06/wwccpn-the-global-operating-environment (accessed on 3 April 2021).
11. Yao, Z.T.; Ji, X.S.; Sarker, P.K.; Tang, J.H.; Ge, L.Q.; Xia, M.S.; Xi, Y.Q. A comprehensive review on the applications of coal fly ash. *Earth-Sci. Rev.* **2015**, *141*, 105–121. [CrossRef]
12. European Coal Combustion Products Association. Statistics. 2016. Available online: http://www.ecoba.com/ecobaccpprod.html (accessed on 3 April 2021).
13. Statistics Poland. Environment 2017. Warsaw. 2017. Available online: https://stat.gov.pl/en/topics/environment-energy/environment/environment-2017,1,9.html (accessed on 9 November 2021).
14. Statistics Poland. Environment 2019. Warsaw. 2019. Available online: https://stat.gov.pl/en/topics/environment-energy/environment/environment-2019,1,11.html (accessed on 9 November 2021).
15. U.S. Energy Information Administration. How Much Carbon Dioxide Is Produced When Different Fuels Are Burned? Available online: https://www.eia.gov/tools/faqs/faq.php?id=73&t=11 (accessed on 7 January 2022).
16. Ritchie, H.; Roser, M. CO_2 and Greenhouse Gas Emissions. Published Online at OurWorldInData.org. 2020. Available online: https://ourworldindata.org/co2-and-other-greenhouse-gas-emissions (accessed on 7 January 2022).
17. Thomas, M.D.A. *Optimizing the Use of Fly Ash in Concrete*; PCA Report IS548; Portland Cement Association: Skokie, IL, USA, 2007.
18. Shaheen, M.S.; Hooda, P.S.; Tsadilas, C.D. Opportunities and challenges in the use of coal fly ash for soil improvements—A review. *J. Environ. Manag.* **2014**, *145*, 249–267. [CrossRef]
19. Wesche, K. (Ed.) *Fly Ash in Concrete: Properties and Performance*; CRC Press: London, UK, 1991.
20. Arvaniti, E.C.; Juenger, M.C.G.; Bernal, S.A.; Duchesne, J.; Courard, L.; Leroy, S.; De Belie, N.; Provis, J.L.; Klemm, A. Determination of particle size, surface area, and shape of supplementary cementitious materials by different techniques. *Mater. Struct.* **2015**, *48*, 3687–3701. [CrossRef]
21. Hosseini Asl, S.M.; Javadian, H.; Khavarpour, M.; Belviso, C.; Taghavi, M.; Maghsudi, M. Porous adsorbents derived from coal fly ash as cost-effective and environmentally-friendly sources of aluminosilicate for sequestration of aqueous and gaseous pollutants: A review. *J. Clean. Prod.* **2019**, *208*, 1131–1147. [CrossRef]

22. Basu, M.; Pande, M.; Bhadoria, P.B.S.; Mahapatra, S.C. Potential fly ash utilization in agriculture: A global review. *Prog. Nat. Sci.* **2009**, *19*, 1173–1186. [CrossRef]
23. Sahu, S.K.; Bhangare, R.C.; Ajmal, P.Y.; Sharma, S.; Pandit, G.G.; Puranik, V.D. Characterization and quantification of persistent organic pollutants in fly ash from coal fueled thermal power stations in India. *Microchem. J.* **2009**, *92*, 92–96. [CrossRef]
24. Zierold, K.M.; Odoh, C. A review on fly ash from coal-fired power plants: Chemical composition, regulations, and health evidence. *Rev. Environ. Health* **2020**, *35*, 401–418. [CrossRef] [PubMed]
25. Rowe, C.L.; Hopkins, W.A.; Congdon, J.D. Ecotoxicological implications of aquatic disposal of coal combustion residues in the United States: A review. *Environ. Monit. Assess.* **2002**, *80*, 207–276. [CrossRef]
26. United States Environmental Protection Agency. Hazardous and solid waste management system; identification and listing of special wastes; disposal of coal combustion residuals from electric utilities; proposed rule (Codified at 40 CFR Parts 257, 261, 264, 265, 268, 271 and 302). *Fed. Regist.* **2010**, *75*, 35127–35264.
27. Singh, R.K.; Gupta, N.C.; Guha, B.K. Fly Ash Disposal in Ash Ponds: A Threat to Ground Water Contamination. *J. Inst. Eng. India Ser. A* **2016**, *97*, 255–260. [CrossRef]
28. Izquierdo, M.; Querol, X. Leaching behaviour of elements from coal combustion fly ash: An overview. *Int. J. Coal Geol.* **2012**, *94*, 54–66. [CrossRef]
29. Nayak, A.K.; Raja, R.; Rao, K.S.; Shukla, A.K.; Mohanty, S.; Shahid, M.; Swain, C.K. Effect of fly ash application on soil microbial response and heavy metal accumulation in soil and rice plant. *Ecotoxicol. Environ. Saf.* **2014**, *114*, 257–262. [CrossRef] [PubMed]
30. Singh, P.K.; Tripathi, P.; Dwivedi, S.; Awasthi, S.; Shri, M.; Chakrabarty, D.; Tripathi, R.D. Fly-ash augmented soil enhances heavy metal accumulation and phytotoxicity in rice (*Oryza sativa* L.); A concern for fly-ash amendments in agriculture sector. *Plant Growth Regul.* **2016**, *78*, 21–30. [CrossRef]
31. Jambhulkar, H.P.; Shaikh, S.M.S.; Kumar, M.S. Fly ash toxicity, emerging issues and possible implications for its exploitation in agriculture; Indian scenario: A review. *Chemosphere* **2018**, *213*, 333–344. [CrossRef]
32. Ahmaruzzaman, M. A review on the utilization of fly ash. *Prog. Energy Combust. Sci.* **2010**, *36*, 327–363. [CrossRef]
33. Komljenović, M.; Petrašinović-Stojkanović, L.; Baščarević, Z.; Jovanović, N.; Rosić, A. Fly ash as the potential raw mixture component for Portland cement clinker synthesis. *J. Therm. Anal. Calorim.* **2009**, *96*, 363–368. [CrossRef]
34. Lechtenbergy, D. Power Plant Ashes as Substitute Raw Materials. Global Cement Magazine. 2012. Available online: https://www.globalcement.com/magazine/articles/735-power-plant-ashes-as-substitute-raw-materials (accessed on 5 April 2021).
35. Kourti, I.; Delgado Sancho, L.; Schorcht, F.; Roudier, S.; Scalet, B.M.; Joint Research Centre, Institute for Prospective Technological Studies. Best Available Techniques (BAT) Reference Document for the Production of Cement, Lime and Magnesium Oxide: Industrial Emissions Directive 2010/75/EU (Integrated Pollution Prevention and Control). Publications Office. 2013. Available online: https://data.europa.eu/doi/10.2788/12850 (accessed on 7 January 2022).
36. Cembureau The European Cement Association. Activity Report, Brussels. 2020. Available online: http://www.cembureau.eu/library/reports/ (accessed on 7 January 2022).
37. *EN 450-1:2012*; Fly Ash for Concrete. Definition, Specifications and Conformity Criteria. British Standards Institution: London, UK, 2012.
38. *EN 197-1:2011*; Cement–Part 1: Composition, Specifications and Conformity Criteria for Common Cements. British Standards Institution: London, UK, 2011.
39. ACI Committee 232. *Use of Fly Ash in Concrete*; ACI 232.2R-96; American Concrete Institute: Farmington Hills, MC, USA, 1996.
40. Mehta, P.K.; Monteiro, P.J. *Concrete: Microstructure, Properties and Materials*, 2nd ed.; McGraw-Hill: New York, NY, USA, 2006.
41. Hashmi, A.F.; Shariq, M.; Baqi, A. An investigation into age-dependent strength, elastic modulus and deflection of low calcium fly ash concrete for sustainable construction. *Constr. Build. Mater.* **2021**, *283*, 122772. [CrossRef]
42. Hashmi, A.F.; Shariq, M.; Baqi, A.; Haq, M. Optimization of fly ash concrete mix—A solution for sustainable development. *Mater. Today Proc.* **2020**, *26*, 3250–3256. [CrossRef]
43. Elsageer, M.; Millard, S.; Barnett, S.J. Strength development of concrete containing coal fly ash under different curing temperature conditions. In Proceedings of the World of Coal Ash (WOCA) Conference, Lexington, KY, USA, 4–7 May 2009.
44. Dong, P.S.; Tuan, N.V.; Thanh, L.T.; Thang, N.C.; Cu, V.H.; Mun, J.-H. Compressive Strength Development of High-Volume Fly Ash Ultra-High-Performance Concrete under Heat Curing Condition with Time. *Appl. Sci.* **2020**, *10*, 7107. [CrossRef]
45. Jitchaiyaphum, K.; Sinsiri, T.; Chindaprasirt, P. Cellular Lightweight Concrete Containing Pozzolan Materials. *Procedia Eng.* **2011**, *14*, 1157–1164. [CrossRef]
46. Liu, X.; Ni, C.; Ji, H.; Tan, S.; Hong, B. Construction Techniques and Quality Test and Evaluation of Lightweight Cellular Concrete Mixed with Fly Ash as Subgrade Material. *Adv. Mater. Sci. Eng.* **2019**, *2019*, 5402679. [CrossRef]
47. Sasovsky, T.A.; Chorna, I.V.; Shalay, S.V.; Lysiak, O.M. Methods of cellular concrete production using fly ash. *Bull. Odessa State Acad. Civ. Eng. Archit.* **2021**, *82*, 114–122. [CrossRef]
48. Zorić, D.; Lazar, D.; Rudić, O.; Radeka, M.; Ranogajec, J.; Hiršenberger, H. Thermal conductivity of lightweight aggregate based on coal fly ash. *J. Therm. Anal. Calorim.* **2012**, *110*, 489–495. [CrossRef]
49. Nadesan, M.S.; Dinakar, P. Mix design and properties of fly ash waste lightweight aggregates in structural lightweight concrete. *Case Stud. Constr. Mater.* **2017**, *7*, 336–347. [CrossRef]
50. Little, M.R.; Adell, V.; Boccaccini, A.R.; Cheeseman, C.R. Production of novel ceramic materials from coal fly ash and metal finishing wastes. *Resour. Conserv. Recycl.* **2008**, *52*, 1329–1335. [CrossRef]

51. Shen, B.; Poulsen, B.; Xun, L.; Qin, J.; Thiruvenkatachiru, R.; Yi, D. Remediation and Monitoring of Abandoned Mines. *Int. J. Min. Sci. Technol.* **2017**, *27*, 803–811. [CrossRef]
52. Strozik, G. The use of Fly Ash for Filling of Shallow Underground Ore Mine Workings on the Example of Mine Area Reclamation in Piekary Śląskie. *Gospod. Surowcami Miner.–Miner. Resour. Manag.* **2018**, *34*, 139–154. [CrossRef]
53. vom Berg, W.; Feuerborn, H.-J. Coal Combustion Products in Europe—Valuable raw materials for the construction industry. *Concr. Plant Int.* **2006**, *4*, 54–62.
54. Directive 2004/42/CE of the European Parliament and of the Council of 21 April 2004 on the Limitation of Emissions of Volatile Organic Compounds Due to the Use of Organic Solvents in Certain Paints and Varnishes and Vehicle Refinishing Products and Amending Directive 1999/13/EC. Available online: https://eur-lex.europa.eu/legal-content/EN/TXT/HTML/?uri=CELEX:32004L0042&from=PL (accessed on 9 November 2021).
55. Code of Federal Regulations 40: Protection of Environment, Chapter 1: Environmental Protection Agency, Subchapter C: Air Programs, Part 51: Requirements for Preparation, Adoption, and Submittal of Implementation Plans, Subpart F: Procedural Requirements. Available online: https://www.ecfr.gov/cgi-bin/text-idx?SID=387a1dbfd3f26c488ff218f51bc0dc3f&mc=true&tpl=/ecfrbrowse/Title40/40CIsubchapC.tpl (accessed on 19 February 2020).
56. Nasini, L.; De Luca, G.; Ricci, A.; Ortolani, F.; Caselli, A.; Massaccesi, L.; Regni, L.; Gigliotti, G.; Proietti, P. Gas emissions during olive mill waste composting under static pile conditions. *Int. Biodeterior. Biodegrad.* **2016**, *107*, 70–76. [CrossRef]
57. US EPA. Initial List of Hazardous Air Pollutants with Modifications. Available online: https://www.epa.gov/haps/initial-list-hazardous-air-pollutants-modifications (accessed on 21 February 2020).
58. Benzene. *IARC Monographs on the Evaluation of Carcinogenic Risks to Humans*; International Agency for Research on Cancer: Lyon, France, 2017; Volume 120. Available online: https://publications.iarc.fr/Book-And-Report-Series/Iarc-Monographs-On-The-Identification-Of-Carcinogenic-Hazards-To-Humans/Benzene-2018 (accessed on 21 February 2020).
59. Directive 2009/30/EC of the European Parliament and of the Council of 23 April 2009 Amending Directive 98/70/EC as Regards the Specification of Petrol, Diesel and Gas-Oil and Introducing a Mechanism to Monitor and Reduce Greenhouse Gas Emissions and Amending Council Directive 1999/32/EC as Regards the Specification of Fuel Used by Inland Waterway Vessels and Repealing Directive 93/12/EEC. Available online: https://eur-lex.europa.eu/legal-content/EN/TXT/HTML/?uri=CELEX:32009L0030&from=PL (accessed on 9 November 2021).
60. Gross, S.A.; Avens, H.J.; Banducci, A.M.; Sahmel, J.; Panko, J.M.; Tvermoes, B.E. Analysis of BTEX groundwater concentrations from surface spills associated with hydraulic fracturing operations. *J. Air Waste Manag. Assoc.* **2013**, *63*, 424–432. [CrossRef] [PubMed]
61. Farshad, A.; Oliaei, H.K.; Mirkazemi, R.; Bakand, S. Risk assessment of benzene, toluene, ethyl benzene and xylenes (BTEX) in pain plants of two automotive industries in Iran by using the coshh guideline. *Eur. Sci. J.* **2013**, *9*, 270–276. [CrossRef]
62. Masekameni, M.D.; Moolla, R.; Gulumian, M.; Brouwer, D. Risk Assessment of Benzene, Toluene, Ethyl Benzene, and Xylene Concentrations from the Combustion of Coal in a Controlled Laboratory Environment. *Int. J. Environ. Res. Public Health* **2019**, *16*, 95. [CrossRef] [PubMed]
63. Cuiting, Y.; Guang, M.; Yunhong, P.; Qibin, X.; Junliang, W.; Zhong, L.; Jing, X. Abatement of various types of VOCs by adsorption/catalytic oxidation: A review. *Chem. Eng. J.* **2019**, *370*, 1128–1153. [CrossRef]
64. Miri, M.; Rostami Aghdam Shendi, M.; Ghaffari, H.R.; Ebrahimi Aval, H.; Ahmadi, E.; Taban, E.; Gholizadeh, A.; Yazdani Aval, M.; Mohammadi, A.; Azari, A. Investigation of outdoor BTEX: Concentration, variations, sources, spatial distribution, and risk assessment. *Chemosphere* **2016**, *163*, 601–609. [CrossRef] [PubMed]
65. Dehghani, M.; Fazlzadeh, M.; Sorooshian, A.; Tabatabaee, H.R.; Miri, M.; Baghani, A.N.; Delikhoon, M.; Mahvi, A.H.; Rashidi, M. Characteristics and health effects of BTEX in a hot spot for urban pollution. *Ecotoxicol. Environ. Saf.* **2018**, *155*, 133–143. [CrossRef] [PubMed]
66. Andreoni, V.; Gianfreda, L. Bioremediation and monitoring of aromatic-polluted habitats. *Appl. Microbiol. Biotechnol.* **2007**, *76*, 287–308. [CrossRef]
67. Zhu, C.; Jia, X.; Zhang, P.; Ye, J.; Wang, H. Cement Stabilization/Solidification of Heavy Metal-Contaminated Sediments Aided by Coal Fly Ash. *Glob. Environ. Eng.* **2017**, *4*, 46–54. Available online: https://www.avantipublishers.com/jms/index.php/tgevnie/article/view/937 (accessed on 9 January 2022).
68. Paria, S.; Yuet, P.K. Solidification–stabilization of organic and inorganic contaminants using portland cement: A literature review. *Environ. Rev.* **2006**, *14*, 217–255. [CrossRef]
69. Hauashdh, A.; Mohamed, R.M.S.R.; Jailani, J.; Rahman, J.A. Stabilization of Peat Soil Using Fly Ash, Bottom Ash and Portland Cement: Soil Improvement and Coal Ash Waste Reduction Approach. *IOP Conf. Ser. Earth Environ. Sci.* **2020**, *498*, 012011. [CrossRef]
70. Blackman, W.C. *Basic Hazardous Waste Management*, 3rd ed.; CRC Press: Boca Raton, FL, USA, 2001.
71. Natali Sora, I.; Pelosato, R.; Botta, D.; Dotelli, G. Chemistry and microstructure of cement pastes admixed with organic liquids. *J. Eur. Ceram. Soc.* **2002**, *22*, 1463–1473. [CrossRef]
72. Karamalidis, A.K.; dan Voundrias, E.A. Cement-based Stabilization/solidification of Oil Refinery Sludge Leaching Behavior of Alkanes and PAHs. *J. Hazard. Mater.* **2007**, *148*, 122–135. [CrossRef] [PubMed]
73. Vipulanandan, C. Effect of Clays and Cement on the Solidification/Stabilisation of Phenol-Contaminated Soils. *Waste Manag.* **1995**, *15*, 399–406. [CrossRef]

74. Butler, L.G.; Owens, J.W.; Cartledge, F.K.; Kurtz, R.L.; Byerly, G.R.; Wales, A.J.; Bryant, P.L.; Emery, E.F.; Dowd, B.; Xie, X. Synchrotron X-ray Microtomography, Electron Probe Microanalysis, and NMR of Toluene Waste in Cement. *Environ. Sci. Technol.* **2000**, *34*, 3269–3275. [CrossRef]
75. *PN-B-04101:1984*; Oznaczanie Nasiąkliwości Wodą. Polski Komitet Normalizacyjny/The Polish Committee for Standardization: Warsaw, Poland, 1984.
76. Banaszkiewicz, K.; Marcinkowski, T. Cementation as a method of remediation of soil contaminated with xylene. *Environ. Prot. Eng.* **2014**, *40*, 57–66. [CrossRef]
77. Banaszkiewicz, K.; Badura, M. Experimental investigation on the application of recycled tires polymer fibers as a BTEX removal material. *SN Appl. Sci.* **2019**, *1*, 558. [CrossRef]
78. Senthil Kumar, M.; Sivasankar, V.; Gopalakrishna, G.V.T. Quantification of benzene in groundwater sources and risk analysis in a popular South Indian Pilgrimage City—A GIS based approach. *Arab. J. Chem.* **2017**, *10*, S2523–S2533. [CrossRef]
79. Logeshwaran, P.; Megharaj, M.; Chadalavada, S.; Bowman, M.; Naidu, R. Petroleum hydrocarbons (PH) in groundwater aquifers: An overview of environmental fate, toxicity, microbial degradation and risk-based remediation approaches Environmental. *Technol. Innov.* **2018**, *10*, 175–193. [CrossRef]
80. ATSDR. *Toxicological Profile for Benzene (Update)*; U.S. Department of Health and Human Services, Public Health Service, Agency for Toxic Substances and Disease Registry: Atlanta, GA, USA, 2007.
81. Lake, C.B.; Arefi, G.; Yuet, P.K. Examining fly ash as a sorbent for benzene, trichloroethylene, and ethylbenzene in cement-treated soils. *Can. Geotech. J.* **2013**, *50*, 423–434. [CrossRef]
82. Sun, J.; Lin, G.; Zhu, H.; Tang, X.; Zhang, L. Study on Adsorption-Desorption of Benzene in Soil. *IOP Conf. Ser. Earth Environ. Sci.* **2020**, *546*, 042041. [CrossRef]
83. Donaldson, S.G.; Miller, G.C.; Miller, W.W. Remediation of Gasoline-Contaminated Soil by Passive Volatilization. *J. Environ. Qual.* **1992**, *21*, 94–102. [CrossRef]
84. Smith, J.A.; Chiou, C.T.; Kammer, J.A.; Kile, D.E. Effect of soil moisture on the sorption of trichloroethene vapor to vadose-zone soil at Picatinny Arsenal, New Jersey. *Environ. Sci. Technol.* **1990**, *24*, 676–683. [CrossRef]
85. Bartoňová, L. Unburned carbon from coal combustion ash: An overview. *Fuel Processing Technol.* **2015**, *134*, 136–158. [CrossRef]
86. Zhao, A.; Pudasainee, D.; Duan, Y.; Gupta, R.; Liu, M.; Lu, J. A review on mercury in coal combustion process: Content and occurrence forms in coal, transformation, sampling methods, emission and control technologies. *Prog. Energy Combust. Sci.* **2019**, *73*, 26–64. [CrossRef]
87. Banaszkiewicz, K.; Marcinkowski, T. Use of cement-fly ash-based stabilization techniques for the treatment of waste containing aromatic contaminants. *E3S Web Conf.* **2017**, *22*, 00009. [CrossRef]
88. Odler, I.; Rößler, M. Investigations on the relationship between porosity, structure and strength of hydrated Portland cement pastes. II. Effect of pore structure and of degree of hydration. *Cem. Concr. Res.* **1985**, *15*, 401–410. [CrossRef]
89. Ezeldin, A.S.; Mikhail, R.; Choi, B. Properties of concrete containing benzene contaminated soil. *ACI Mater. J.* **1995**, *92*, 401–410.
90. Rafieizonooz, M.; Mirza, J.; Salim, M.R.; Hussin, M.W.; Khankhaje, E. Investigation of coal bottom ash and fly ash in concrete as replacement for sand and cement. *Constr. Build. Mater.* **2016**, *116*, 15–24. [CrossRef]
91. Saha, A.K. Effect of class F fly ash on the durability properties of concrete. *Sustain. Environ. Res.* **2018**, *28*, 25–31. [CrossRef]
92. Chen, X.; Wu, S. Influence of water-to-cement ratio and curing period on pore structure of cement mortar. *Constr. Build. Mater.* **2013**, *30*, 804–812. [CrossRef]
93. Siline, M.; Ghorbel, E.; Bibi, M. Effect of freeze-thaw cycles on the physicomechanical properties of a pozzolanic mortar. *Constr. Build. Mater.* **2017**, *134*, 32–38. [CrossRef]

Article

The Effect of Ash Silanization on the Selected Properties of Rigid Polyurethane Foam/Coal Fly Ash Composites

Beata Zygmunt-Kowalska [1], Kinga Pielichowska [2], Patrycja Trestka [1], Magdalena Ziąbka [3] and Monika Kuźnia [1,*]

[1] Department of Heat Engineering and Environment Protection, Faculty of Metals Engineering and Industrial Computer Science, AGH University of Science and Technology, Mickiewicza 30 Av., 30-059 Krakow, Poland; zygmunt@agh.edu.pl (B.Z.-K.); trestka@agh.edu.pl (P.T.)
[2] Department of Biomaterials and Composites, Faculty of Materials Science and Ceramics, AGH University of Science and Technology, Mickiewicza 30 Av., 30-059 Krakow, Poland; kingapie@agh.edu.pl
[3] Department of Ceramics and Refractories, Faculty of Materials Science and Ceramics, AGH University of Science and Technology, Mickiewicza 30 Av., 30-059 Krakow, Poland; ziabka@agh.edu.pl
* Correspondence: kuznia@agh.edu.pl; Tel.: +48-12-617-31-98

Citation: Zygmunt-Kowalska, B.; Pielichowska, K.; Trestka, P.; Ziąbka, M.; Kuźnia, M. The Effect of Ash Silanization on the Selected Properties of Rigid Polyurethane Foam/Coal Fly Ash Composites. *Energies* **2022**, *15*, 2014. https://doi.org/10.3390/en15062014

Academic Editor: Konstantinos S. Triantafyllidis

Received: 30 January 2022
Accepted: 4 March 2022
Published: 10 March 2022

Publisher's Note: MDPI stays neutral with regard to jurisdictional claims in published maps and institutional affiliations.

Copyright: © 2022 by the authors. Licensee MDPI, Basel, Switzerland. This article is an open access article distributed under the terms and conditions of the Creative Commons Attribution (CC BY) license (https://creativecommons.org/licenses/by/4.0/).

Abstract: According to the assumptions of the European Union, by 2050 it is planned to achieve climate neutrality. For this purpose, a document called the "European Green Deal" was established, which is a set of policies of the European Commission. One of the assumptions is a circular economy that takes into account the use of waste in subsequent production cycles. In order to meet the latest trends in environmentally friendly materials and use of waste in the production of building materials, composites of rigid polyurethane foam with 10 wt.% of waste were produced. Fly ash from coal combustion after modification was used as a filler. Three types of modifications were used: silanization, sieving, and both processes together. The silanization process was carried out for 1 and 2% silane ([3-(2-aminoethylamino)propyl]trimethoxysilane) concentration in relation to the fly ash mass. The sieving was aimed at reaching a fraction with a particle diameter below 75 μm. Six composites with modified fillers were compared and one material containing unchanged fly ash was used as a reference. A comparative analysis was carried out on the basis of surface analysis, thermal stability and physical properties. It turned out that the polyurethane materials modified fly ash silanized with 1% and 2% silane solution proved the best results in performed tests. On the other hand, the polyurethane foam containing sieved ash was characterized by the lowest flammability and the lowest emission of smoke and CO. The use of modified fly ash in technology of polyurethane foams can be a good method of its disposal and can increase the applicability of the composites.

Keywords: fly ash recycling; silanization; polyurethane composites

1. Introduction

Currently, all over the world there is a trend of creating environmentally friendly polymer materials with a reduced amount of petrochemical products, often replaced with plant substrates or fillers [1]. The majority of fillers have got hydrophilic nature. It causes low adhesion between filler and polymer matrix [2]. The use of fillers may disturb the structure of the material and deteriorate the properties of the material. To increase the compatibility between a modifier and a polymer matrix, modifications of the filler are used. The following methods are used in the literature: alkalization (change of surface topography with roughness) and silanization (introducing functional groups to fillers). After the silanization process, the silane molecule acts as a coupling agent to form a chemical bond between the solid modifier and the polymer matrix [3]. This modification method is successfully used in dental composites [4], bone cement formulations [5], esophageal stents [6]. Silanization has not only been widely used in medicine, but also in engineering.

An example could be a silanized filler in resin composites [7,8], polyurethane foams [9,10] and in epoxy-based laminates [11].

In this work, fly ash from coal combustion was subjected to the silanization process. In the world literature, there are articles whose purpose was to modify the fly ash. Goh et al. [12] added silanized fly ash to the epoxy resin and on the basis of the results of the SEM analysis and strength tests showed that the modification of the additive surface increased the filler-polymer interfaces. Şen et al. [13] showed that the silanized filler did not shorten the curing time of the resin and that the silanization of the fly ash reduced the swelling. What is more silanized fly ash has been used in materials such as poly(vinyl chloride) composites [14] and in geopolymer [15]. In the available literature, no studies on the use of silanized fly ash in the rigid polyurethane foam technology were found. In this paper, the possibilities of filler modification that could improve the properties of the composite were investigated. Due to the necessity to seek for new environmentally neutral flame retardants for polymers, ex-tensive research was carried out on the influence of fly ash on the flammability of the material. There are results available in the literature on unmodified FA in PU foam, but none of them concern flammability.

Rigid polyurethane foams (RPUFs) are a group of closed-cell polymers whose structure is characterized by high cross-linking. RPUFs account for 23% of all polyurethane production in the world [16,17]. RPUFs have good thermal insulation properties, low apparent density and good mechanical properties. Their advantages also include resistance to weather conditions. Due to all these features, RPUFs are often used in everyday life, incl. construction, furniture and vehicles [18–20]. The most important application is in construction, as thermal insulation materials. According to the data, buildings are responsible for over 1/3 of the total energy consumption in the world [21]. Due to this fact, it is crucial to use energy in more efficient way. The demand for RPUFs is constantly growing, which implies that this material is more and more widely researched and described in the literature.

The polyurethane (PU) industry is strongly related to petrochemical products. For this reason, polyurethane foams are considered as less environmentally friendly materials. Additionally, RPUFs are more expensive compared to other thermal insulation materials. Therefore, in the PU industry inorganic and organic fillers are often used [22]. The use of fillers improve the mechanical properties, have an influence on the flammability and other RPUF properties. One of the inorganic fillers described in the literature is silica. This filler improves the mechanical properties and reduces the flammability of RPUFs [23,24]. Another type of filler that reduces flammability and, on the other hand, deteriorates mechanical properties is expanded graphite [25]. In PU foams, in order to strengthen them, inorganic fillers are used carbon fiber, PET, carbon nanotube, nanoclay and talc [26–29]. On the other hand, organic fillers are increasingly described in the literature. One of the limited fillers used in the synthesis of RPUFs are cellulose nanocrystals. This filler improves the mechanical properties, dimensional stability and reduces the absorbability of the obtained composites [30]. Egg shells, walnut shells or chitin extracted from arthropods are other fillers that modify the properties of rigid foams PU. The conducted research has shown that a small addition of this type of fillers has a positive effect on the mechanical properties of the foams [31].

In this article, the effect of modification (silanization and sieving) on fly ash from coal combustion on polyurethane composites was evaluated. The evaluation was based on the research results of thermal, mechanical and flammability properties.

2. Materials and Methods

2.1. Characterization of Filler and Its Modification

Fly ash (FA) from the combustion of coal is used as a filler. The fly ash came from a conventional (pulverized) hard coal combustion boiler from power plant in southern Poland. The ash was collected from the electrostatic precipitators. The physical density of the fly ash is 2.15 g/cm^3. According to data provided from manufacturer the ash has category A (taking into account the loss on ignition). The filler particles size are varied from a few to about 500 μm. In this work, fly ash was introduced into polyurethane in an unmodified form and after the silanization and sieving process. Conventional fly ash is characterized by the presence of unburned porous carbon particles which could affect the flammability of the material. Therefore, it was decided to sift the filler on a sieve with a mesh diameter of 75 μm. The fly ash was modified by sieving and silanizing it. In earlier work, the authors investigated the effect of particle size on their chemical composition [32]. It turned out that above the particle diameter of 75–100 μm, the amount of unburned carbon particles increased significantly. Therefore, it was decided to modify the fly ash, which consisted in screening and separating the fraction below 75 μm. The second part of modification was silanization, the main purpose of which was to increase the adhesion of the particles. On the other hand, the chemical composition of fly ash depending on the particle sizes (apart from carbon content) practically does not differ from each other (before and after sieving). Fly ash consists mainly of: SiO_2 (52.3%), Al_2O_3 (27.4%), Fe_2O_3 (7.9%), K_2O (4.1%), CaO (2.9%), TiO_2 (1.4%) and MgO (1.4%) [33,34]. The analyzed fly ash belongs to the group of silicate fly ash due to the large amount of SiO_2 and in accordance with the ASTM C 618 standard, it is classified as an ash of the group F. Figure 1 shows the structure of fly ash and two characteristics types of particles: microspheres and unburned carbon. These elements occur in pulverized fly ash, they are not formed during the combustion of coal in a fluidized bed. The screened residue can be seen in photo d, EDS analysis showed that this fraction contains large amounts of carbon.

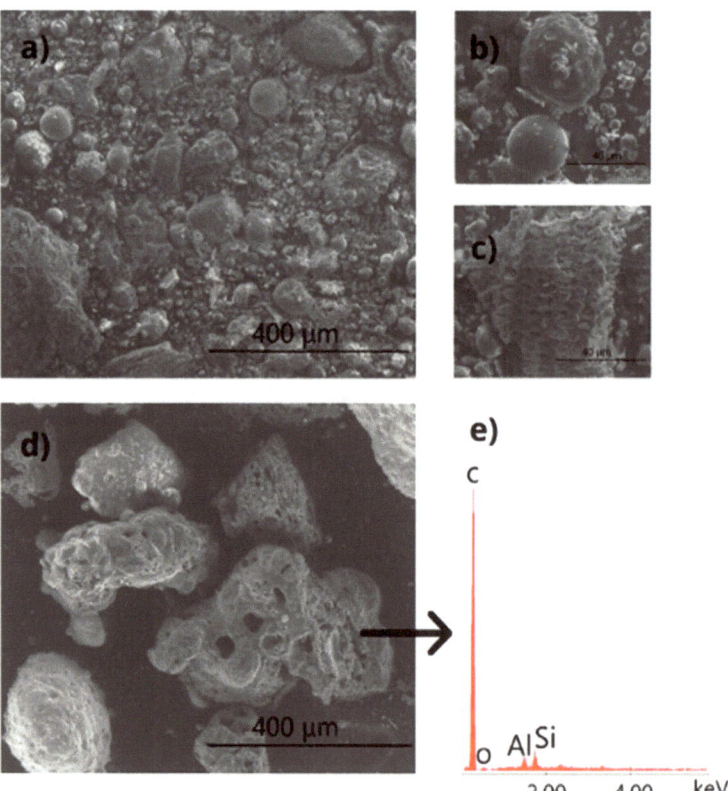

Figure 1. Fly ash microphotography (**a**), microsphere microphotography (**b**), unburned carbon microphotography (**c**), screened residue microphotography (**d**), EDS spectroscopy of residue (**e**).

The next step of the content modification is silanization. Figure 2 shows the scheme of silanization process.

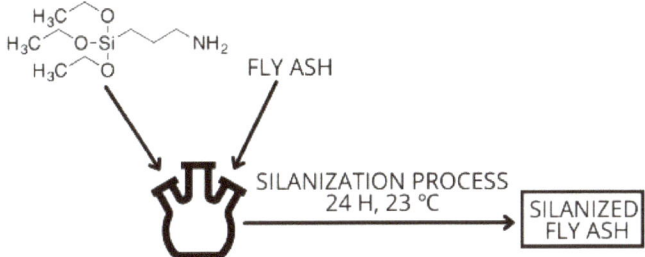

Figure 2. Scheme of silanization process.

This alteration of filler was made by toluene method according to [35] in room temperature (23 °C) and for 24 h. The silanization was performed for various silane concentrations. The weight of silane in filler is 1% (1SFA) and 2% (2SFA) of the weight of fly ash. Additionally fillers were sieved before silanization to obtain two more fillers (1SPFA, 2SPFA).

For all fillers elemental analysis of carbon, nitrogen and hydrogen was performed using LECO CHN628 device. The analysis of particle size of fillers was performed used the methods of laser light diffraction and the Mastesizer 2000S apparatus (Malvern Instruments Ltd., Malvern, UK). The particle size distribution of fillers is shown in Figure 3. It can be seen that the screened fly ash (PFA) and screened and silanized fly ash (1SPFA, 2SPFA) have a particle diameter below 100 μm, therefore it contains much less unburned carbon particles (approx. 6–7%), which is confirmed by elemental analysis. On the other hand, non-sieved (FA) and non-sieved and silanized fly ash (1SFA, 2SFA) has a lot of particles with a diameter of several hundred μm, which is related to the higher content of unburned carbon (approx. 9–10%). The specific surface area (SSA) of samples was determined using the multipoint BET adsorption method (ASAP 2010, Micromeritics Instrument Corporation, Norcross, GA, USA). Explanation of filler symbols and analysis results are presented in the Table 1.

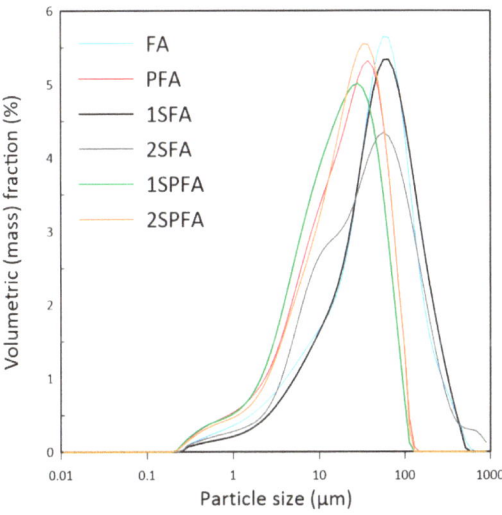

Figure 3. Particle size distribution of fillers.

Table 1. Elemental analysis of C, H, N and SSA of fillers.

Filler	Description	C (%)	H (%)	N (%)	SSA (m^2/g)
FA	unmodified fly ash	9.46	0.00	0.19	5.47
PFA	sifted fly ash	6.93	0.02	0.17	3.23
1SFA	fly ash silanized with 1% silane solution	9.95	0.15	0.30	2.24
2SFA	fly ash silanized with 2% silane solution	10.47	0.19	0.49	0.95
1SPFA	sifted fly ash silanized with 1% silane solution	6.52	0.08	0.27	1.09
2SPFA	sifted fly ash silanized with 2% silane solution	6.42	0.04	0.21	2.21

The sieved fillers, i.e., PFA, 1SPFA, 2SPFA were characterized by the smallest amount of carbon, about 30% less unburned carbon particles compared to the FA filler. The raw fly ash (not sieved and unsilanized) is characterized by the highest value of the SSA, which amounts to over 5 m^2/g. The other values do not create any dependence, therefore the research should be extended in terms of the mechanism of SSA formation.

2.2. Preparation of Modified Polyurethane Foams

The main components of polyurethane rigid foam are additives (10 wt.% of fly ash, modified fly ash), polyol system with blowing agent, fire retardants, catalyst (component A). The last element of the material is isocyanate prepolymer (component B). Polyurethane foams were prepared using the two-component commercial system EKOPRODUR PM4032 (PCC Group, Brzeg Dolny, Poland). Components were mixed at a ratio of 100:110 by weight (according to product characteristic). Firstly the polyol was mixed with the fly ash until the homogeneous mixture was obtained, after that the isocyanate was poured into the mixture and it was mixed together by mechanical agitator for 8 s at 1200 rpm. The well mixed polyurethane-fly ash mixture was poured into square mold. The composites were left in the forms for 24 h. After this time, the foams were removed from the mold and prepared for tests. The production process of polyurethane composite modified with fly ash is presented in the Figure 4.

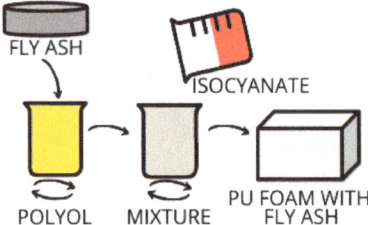

Figure 4. Schematic presentation of the production of polyurethane composites.

2.3. Methods

The foaming process. The foaming process was analyzed by determining characteristic times: cream time–the time from mixing components A and B to the beginning of the increase in the volume of the resulting mixture, rise time–the time after which it is achieved there is a maximum increase in RPUFs, tack free time–time after which a non-sticky layer form on the RPUFs surface.

Microstructure. The RPUFs microstructure was observed with the scanning electronic microscope (SEM; Nova NanoSEM 200; FEI Company, Hillsboro, OR, USA). Samples of dimensions 50 × 50 × 50 mm^3 were sputtered with gold and analyzed at an acceleration of 10 kV.

Thermal analysis. Thermogravimetric (TG) analysis of the obtained materials was performed on a Discovery TGA 550 (TA Instruments, New Castle, DE, USA) to evaluate the thermal properties of the obtained RPUFs. The thermal behavior of a samples (ca. 10 mg) was analyzed in platinum pans under nitrogen atmosphere (20 mL/min) and in the temperature range 20–700 °C at a heating rate of 10 °C min^{-1}.

Density. The density for RPUFs was determined in accordance with the standard EN ISO 845:2009. The sample used for the tests had a volume of 100 cm^3.

Friability. Friability was determined during an experiment carried out according to the ASTM C421-08 standard. Twelve cubic cubes of foam were weighted. Then they were rotated with the oak cubes in an oak box for 10 min at speed 60 r/min. After the process, the foam samples were reweighted.

Dimensional stability. The dimensional stability for RPUF at 70 °C was determined after 48 h. The measurement was carried out in accordance with the ASTM D2126-09 standard. The change in length was calculated.

Water absorption. Water absorption was calculated according to ASTM D570-98 standard. The samples were immersed in distilled water and their weight was deter-mined after 48 h.

Limited Oxygen Index. The limitation of the oxygen index (LOI) was the test was measured according to ISO 4589-2: 2017. The samples used for the test were 150 × 10 × 10 mm^3.

Heat release intensity. The average heat release rate (HRR), maximum heat release rate (PHRR), effective heat of combustion (EHC), time to ignition (TTI), percent mass loss (PML) of the tested sample were determined on the conical calorimeter. The test was measured according to ISO 5660:2015. The samples used for the test were 30 × 100 × 100 mm^3.

Flammability. Flammability tests were carried out using the UL 94V. A vertical burning test was performed. For the test, samples with dimensions of 125 × 13 × 13 mm^3 were used. The sample was placed in a holder and a burner was located underneath it.

Reaction to fire. Reaction to fire tests of the samples were also carried out by determining the gross calorific value (LECO AC500 isoperibolic calorimeter). The measurement was performed in accordance with the ISO 1716:2018 standard. A sample with a mass of 80 mg placed in a calorimetric bomb was completely burnt in an atmosphere of oxygen under pressure.

3. Results and Discussion

3.1. Foaming Characteristics of Polyurethane Composites

The impact of fillers and their modification on foaming kinetics of composites was assessed by foaming parameters such as cream time, rise time and tack-free time. The obtained samples and characteristic times can be found in the Table 2.

Table 2. Foaming parameters of polyurethane composites.

Foam Code	Cream Time (s)	Rise Time (s)	Tack Free Time (s)
PU_FA	45	236	350
PU_PFA	51	251	330
PU_1SFA	42	281	310
PU_2SFA	45	284	310
PU_1SPFA	49	266	305
PU_2SPFA	50	259	300

Based on the analysis of the results presented in the Table 2, it was found that the modification of the filler affects the kinetics of foam forming. The screened fly ash filler increased the forming time of the foams. This is due to its higher density, which affects the viscosity of the polyol-fly ash mixture. As a result it has a significant impact on foam behavior and increases reaction time [36,37]. For composites containing silanized fillers (1SFA and 2SFA), the rise time is higher compared to others. It is caused by the reaction of the hydroxyl groups present in the filling material after the silanization process, which react quickly with isocyanate groups and influence the correct stoichiometry of the material synthesis [38].

3.2. Properties of Polyurethane Composites

Apparent density is an important parameter that affects the mechanical properties of the RPUFs.

The apparent density presented in Table 3 for PU_FA with unmodified filler was 38.4 [kg·m^{-3}]. The modification of the filler by silanization resulted in a reduction of the apparent density of the foams. This is due to the fact that the filler's silanization process improves its compatibility with the polyurethane matrix. Foams with a higher cross-link density are created. For this reason, the apparent density value for foams with silanized filler decreases [38]. The modification of the filler by sieving and its subsequent silanization had the opposite effect. The PU_PFA, PU_1SPFA and PU_2SPFA foams have a higher apparent density value than the reference foam. Typical RPUFs should have an apparent density in the range of 28–60 [kg·m^{-3}] [39]. The obtained foams have density within the recommended range.

Table 3. Results of apparent density, dimensional stability and water absorption of RPUFs.

Sample	Apparent Density (kg·m^{-3})	Dimensional Stability (Δl, 48 h, 70 °C) (%)	Loss in Mass (Δm, 48 h, 70 °C) (%)	Water Absorption (48 h, 21 °C) (%)
PU_FA	38.4	1.20	0.45	48.68
PU_PFA	39.8	0.40	0.55	76.16
PU_1SFA	37.3	1.40	0.43	45.45
PU_2SFA	37.8	0.70	0.41	45.29
PU_1SPFA	39.8	0.40	0.50	77.92
PU_2SPFA	40.8	0.45	0.54	76.23

The change of linear dimensions (Δl) and weight loss (Δm) of the analyzed RPUFs after conditioning at 70 °C for 48 h is presented in Table 4. Based on the results of accelerated aging tests for the obtained RPUFs with modified fillers, it was observed that the weight loss was in the range of 0.41–0.55%. The PU_2SFA sample had the lowest weight loss. Additionally, the change in linear dimensions did not exceed 1.40%. The PU_PFA and PU_1SPFA samples are the most thermally stable. In general, it can be seen that the results obtained are random and there is no logical cycling in them. However, the described foams comply with the provisions of the building code, which indicate that polyurethane materials subjected to the increased temperature cannot show changes in linear dimensions greater than 3% [22,40].

Table 4. Characteristics of thermal degradation of the PU composites.

Sample	$T_{1\%}$ (°C)	$T_{5\%}$ (°C)	$T_{10\%}$ (°C)	$T_{50\%}$ (°C)	T_{DTGmax} (°C)	Residue at 600 °C (%)
PU_FA	250	276	290	378	305	32.0
PU_PFA	250	277	291	382	308	32.9
PU_1SFA	251	277	291	379	306	31.7
PU_2SFA	248	274	288	372	303	29.6
PU_1SPFA	250	277	291	386	308	33.0
PU_2SPFA	251	278	292	388	307	32.7

In porous materials water absorption is related to the number of closed and open cells in their structure [41]. Table 3 presents the results of water absorption for the RPUF with modified fillers. Foams with silanized filler-PU_1SFA and PU_2SFA are characterized by low water absorption. This is due to the fact that PU_1SFA and PU_2SFA have a more uniform structure and a larger number of cells with smaller diameter. Such cells cannot store large amounts of water [42]. On the contrary, the modification of the filler by sieving causes an increase in water absorption by about 30%. Connection between apparent density and water absorption is presented in the Figure 5. The foams with a higher apparent density were characterized by significantly higher water absorption parameters.

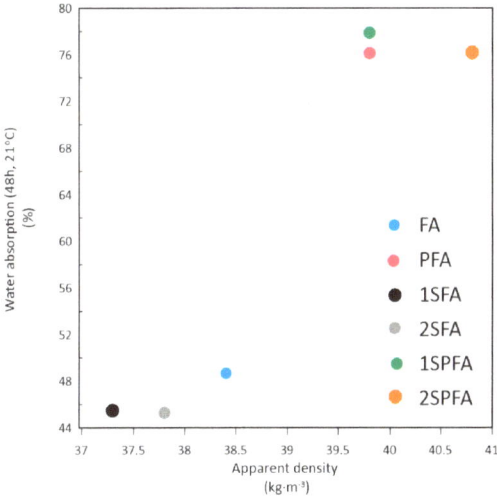

Figure 5. Dependence between apparent density and water absorption of samples.

The friability results for polyurethane foams with modified fillers are depicted in Figure 6.

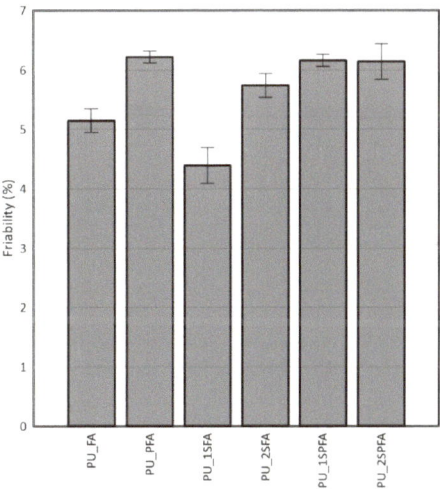

Figure 6. Dependence of the brittleness of polyurethane composites on the type of filler used.

The brittleness of the tested foams was in the range of 4.39–6.22%. The PU_1SFA sample showed the lowest weight loss in the friability test, which indicates a positive effect of FA silanization on the properties of foams. As we mentioned earlier, the mechanical properties, including friability, are closely related to the shape of cells in the foam structure and its apparent density [43]. PU_1SFA analysis indicated that it has the lowest apparent density value among the analyzed materials. The positive effect of introducing the filler into the RPUF was also observed by Barczewski et al. [44]. In contrast, modification of the filler by sieving and silanization increased the friability of the foams.

3.3. Microstructure Analysis of Polyurethane Composites

The influence of filler modification on the porous microstructure of RPUFs was carried out using scanning electron microscopy (SEM). Exemplary images of the morphology of the analyzed materials are presented in the Figure 7.

The morphology of the foam with unmodified filler (PU_FA) depicted in Figure 7a shows that the size and distribution of the cells are heterogeneous. Closed cells were characteristic for the foam structure.

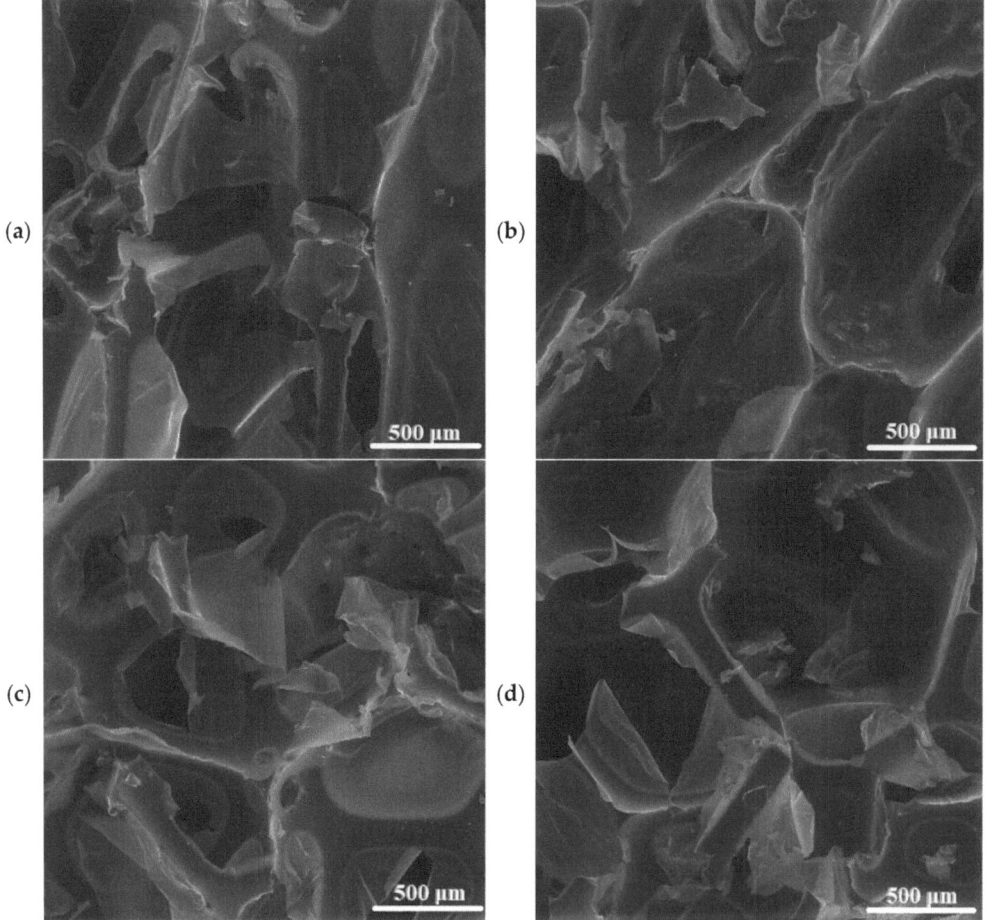

Figure 7. *Cont.*

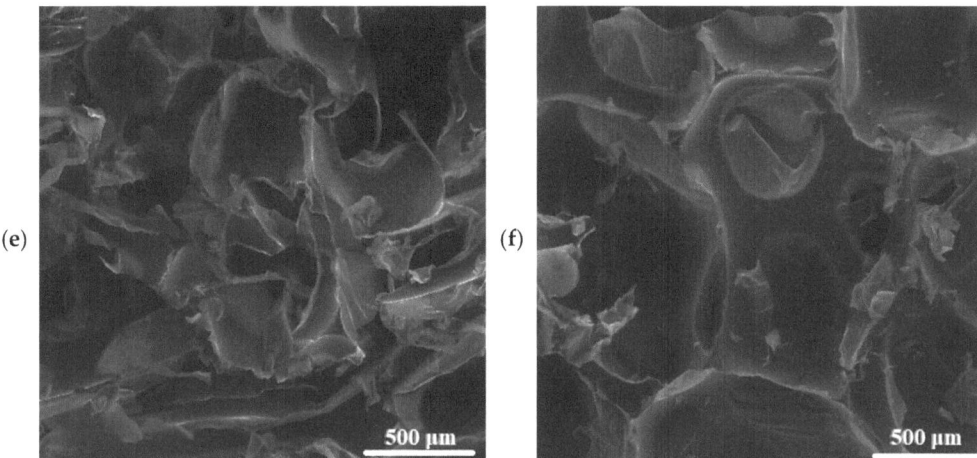

Figure 7. SEM images of the PU_FA (**a**), PU_PFA (**b**), PU_1SFA (**c**), PU_2SFA (**d**), PU_1SPFA (**e**), PU_2SPFA (**f**).

Modification of the filler by sieving has a positive effect on the morphology of RPUFs as the size and distribution of cells are more homogeneous (Figure 7b) compared to the reference foam (Figure 7a). Additionally, modification of the filler by silanization reduces the size of the cells. The analysis of PU_1SPFA and PU_2SPFA morphology showed a greater number of damaged cells.

The morphology of the silanized FA modified RPUFs is shown in Figure 7c,d. The introduction of this type of filler causes the structure to become uniform and the thickness of the cells' walls is reduced. The filler silanization process reflects in more spherical shape of the cells [45]. In the case of PU_2SFA, a greater number of open cells is associated with a reduced adhesion between the polymer matrix and the surface of the filler [29].

3.4. Thermal Properties of Polyurethane Composites

The thermal stability of RPUFs was determined by TG method. Table 4 shows the characteristic temperatures, i.e., with a 5% and 50% weight loss of the tested sample and the char residue after thermal treatment. Polyurethane composites containing modified fillers are less thermally stable than the reference foam (PU_FA). The PU_FA degradation process starts at 194 °C, while for foams with silanized filler it is already at 186 °C, and for foams with sieved and then silanized ash it is already at 191 °C. Filled modified foam, on the other hand, has a lower maximum weight loss than the reference foam, with the exception of PU_2SFA. Partial deterioration of thermal properties may be related to the uneven distribution of the filler in the polymer matrix [1]. The TG analysis showed that foams containing the modified fillers have a higher residue after analysis than PU_FA. Only PU_2SFA had less residual. The faster process of PU_2SFA foam degradation may be influenced by the higher content of open cells (Figure 7d).

As can be seen from the TGA results (Table 4 and Figure 8) incorporation of flay ash-based additives does not change the thermal stability of the polyurethane matrix. The first mass loss is related to the evaporation of moisture and unreacted volatile subtances and unreacted isocyanate monomers that remained in the materials [46]. Themal decomposition of all investigated composites starts above 250 °C. The maximum degradation rate and the selected percentage of mass loss occur at similar temperatures for all PUR composites. As can be seen from DTG curves, degradation process in analyzed materials takes place in two overlapping stages at ca. 250–500 °C. In the first stage, the hard segments of the RPUFs structure are degraded, which is related to the braking of urethane bonds. In the second stage, degradation of the polyol-based soft segments of the RPUFs structure occurs and leads to formation of some kinds of aliphatic ethers or alcohols [46].

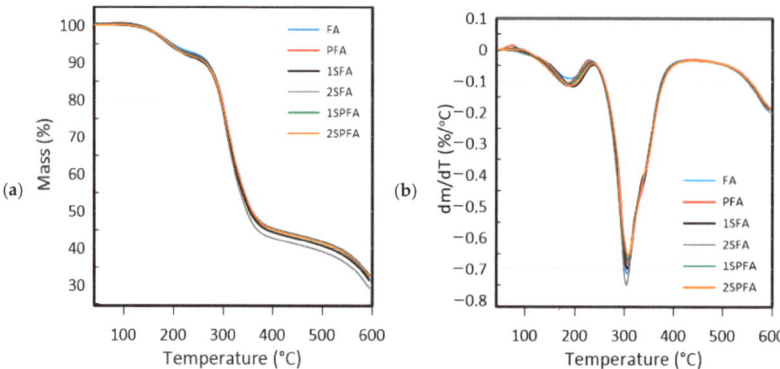

Figure 8. TG (**a**) and DTG (**b**) curves of PUR composites.

3.5. Flammability

The effect of filler modification on the gross calorific value of polyurethane composites was determined using a calorimetric bomb is summarized in Table 5.

Table 5. Gross calorific value, LOI values, UL-94 vertical burning behaviors of polyurethane composites.

Sample	Gross Calorific Value (MJ·kg^{-1})	LOI (%)	UL94
PU_FA	24.4	22.0	N.R.
PU_PFA	23.5	21.8	N.R.
PU_1SFA	23.2	21.8	N.R.
PU_2SFA	23.5	21.9	N.R.
PU_1SPFA	23.7	22.0	N.R.
PU_2SPFA	23.7	22.0	N.R.

The obtained results showed that the modifier treatment influenced the value of the gross calorific value of polyurethane composites. The highest value is shown by the foam that contains unmodified fly ash (24.4 MJ·kg^{-1}) The same value was characteristic for the rigid polyurethane foam containing 10% fly ash addition in other published studies [47]. Nonetheless, the lowest value (23.2 MJ·kg^{-1}) was obtained for sample PU_1SFA. The decrease in the heat of combustion value in the case of PU_FA and PU_1SFA is 5%. This means that, for the same amount of filler in the foam, it is possible to further reduce the heat of combustion by modifying the filler. Sieving also decreased the combustion heat value. Moreover, the Table 5 summarizes the results of LOI and vertical tests performed in accordance with the UL94 standard. The oxygen index values for all samples oscillate around 22%. Meanwhile, all samples have no rating (N.R.) in vertical burning tests. It means that 10% addition of unmodified and modified fly ash is not sufficient to retard the flame. In the literature, no UL94 and LOI results were found for rigid polyurethane foams containing fly ash.

The cone calorimeter tests were carried out to assess the flammability and smoke production. The measuring instrument is effective in assessing the flame retardance of materials as it reflects the fire under real conditions [47]. The Table 6 presents the results of the time to ignition (TTI), effective heat of combustion (AEHC), maximum heat release rate (pHRR), time to pHRR (T-pHRR), fire growth rate (FIGRA), maximum average rate of heat emission (MARHE) and total smoke release (TSR).

Table 6. Cone calorimeter results of polyurethane composites.

Sample	TTI (s)	AEHC (MJ/kg)	pHRR (kW/m^2)	T-pHRR (s)	FIGRA (kW/m^{2}s)	MARHE (kW/m^2)	TSR (m^2/m^2)
PU_FA	2	28.1	207.8	26	7.99	177.16	679.9
PU_PFA	6	18.1	164.1	32	5.13	136.52	579.5
PU_1SFA	6	19.3	178.2	86	2.07	131.35	571.2
PU_2SFA	6	19.0	180.9	44	4.11	144.95	555.9
PU_1SPFA	4	20.6	197.2	70	2.82	151.63	544.5
PU_2SPFA	4	21.8	187.6	32	5.86	157.24	552.2

The shortest TTI time was obtained for PU_FA. The highest ignition delay (4 s. in relation to PU_FA) was achieved for PU_PFA and PU_1SFA samples. The highest values of AEHC and pHRR were found for the foam with unmodified FA and the lowest for foam with sieved fly ash. Sifting out of fly ash of the unburned carbon fraction and reducing the filler particles allowed to reduce the maximum HRR by 21%. Silanization of the modifier allowed to reduce pHRR by 13% in relation to the PU_FA. FIGRA is important index for heat risk assessment. It is defined as the ratio of the maximum HRR (p-HRR) value to the time the peak occurred (T-pHRR) [48]. Due to the FIGRA parameter, PU_FA was characterized by the greatest heat risk. A similar dependence of the results was observed for the MARHE parameter. The TSR decreased with each modification of the foam filler. The best value was obtained for the foam with the filler subjected to both modifications: sieved and silanized (PU_1SPFA).

The Figure 9 shows the charts of heat release rate (HRR), total heat release (THR), CO emission, CO$_2$ emission.

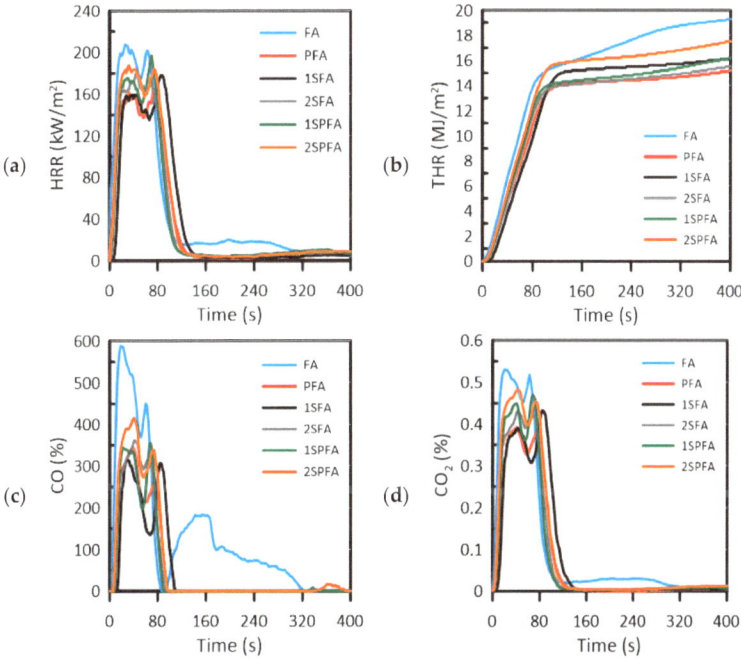

Figure 9. Charts of HRR (**a**), THR (**b**), CO (**c**), CO$_2$ (**d**).

The HRR plots of PU composites have the largest peak with a maximum value of approx. 208 kW/m^2 (Table 6). The lowest peak of the HRR curves was obtained for PU_PFA. On the basis of the analysis of the curves, it was observed that after 160 s to about 300 s, the foam containing unmodified ash continued to emit heat of about 20 kW/m^2. In the case of the HRR curves, the highest value for the TSR curves was obtained for PU_FA and then PU_2SPFA. The lowest value of the curve (maximum approx. 14 MJ/m^2) was determined for the PU_PFA plot. The highest emissions of CO and CO_2 were also obtained for the foam containing unmodified fly ash.

4. Conclusions

In the study, three modifications of the filler (fly ash) were performed: silanization, sieving, silanization and sieving, and the influence of the filler on the properties of polyurethane composites was investigated. The reference foam was a foam containing unmodified coal fly ash.

The foams containing sieving fly ash (PU_PFA, PU_1SPFA, PU_2SPFA) were characterized by higher apparent density, water absorption and friability compared to the other samples. Despite the low values of the above parameters, they were characterized by reduced flammability, the lowest flammability values were obtained for the PU_PFA sample. As a result, the authors proved that a simple modification, which is sieving, reduces the amount of heat released and the emission of CO and CO2 during combustion of polyurethane material.

Samples containing silanized fillers were characterized by the lowest apparent densities and water absorption. The lowest brittleness values of the material were obtained for the PU_1SFA sample. PU_1SFA foam also showed good flammability properties, however, they were still worse than PU_PFA.

The use of fly ash in technology of PUR foam is a great solution that reduces the amount of petrochemical products in the production of polyurethane materials and is a good method of waste disposal. The modifications presented in the article may improve the properties of the fly ash-polyurethane composite, however, before starting the adjustment process, the material properties to be improved should be selected and the appropriate modification method adjusted.

Author Contributions: Conceptualization, B.Z.-K., P.T. and M.K.; methodology, B.Z.-K., P.T., M.K. and K.P.; investigation, B.Z.-K., P.T., M.K., K.P. and M.Z.; resources, B.Z.-K., P.T. and M.K.; writing—original draft preparation, B.Z.-K., P.T. and M.K.; writing—review and editing, B.Z.-K., P.T., M.K. and K.P.; visualization, B.Z.-K. and M.K.; supervision, B.Z.-K. and M.K. All authors have read and agreed to the published version of the manuscript.

Funding: This research was funded by the Ministry of Science and Higher Education, Poland [grant AGH-UST no 16.16.110.663].

Institutional Review Board Statement: Not applicable.

Informed Consent Statement: Not applicable.

Data Availability Statement: Not applicable.

Conflicts of Interest: The authors declare no conflict of interest.

References

1. Hassan, A.A.; Formela, K.; Wang, S. Enhanced interfacial and mechanical performance of styrene-butadiene rubber/silica composites compatibilized by soybean oil derived silanized plasticization. *Compos. Sci. Technol.* **2020**, *197*, 108271. [CrossRef]
2. Masłowski, M.; Miedzianowska, J.; Strzelec, K. Silanized cereal straw as a novel, functional filler of natural rubber biocomposites. *Cellulose* **2019**, *26*, 1025–1040. [CrossRef]
3. Dharmalingam, S.; Meenakshisundaram, O.; Kugarajah, V. Effect of degree of silanization of luffa on the properties of luffa-epoxy composites. *Colloids Surf. A Physicochem. Eng. Asp.* **2020**, *603*, 25273. [CrossRef]

4. Lewis, S.H.; Fugolin, A.P.P.; Lam, S.; Scanlon, C.; Ferracane, J.L.; Pfeifer, C.S. Effects of systematically varied thiourethane-functionalized filler concentration on polymerization behavior and relevant clinical properties of dental composites. *Mater. Des.* **2021**, *197*, 109249. [CrossRef] [PubMed]
5. Cisneros-Pineda, O.G.; Herrera Kao, W.; Loría-Bastarrachea, M.I.; Veranes-Pantoja, Y.; Cauich-Rodríguez, J.V.; Cervantes-Uc, J.M. Towards optimization of the silanization process of hydroxyapatite for its use in bone cement formulations. *Mater. Sci. Eng. C* **2014**, *40*, 157–163. [CrossRef] [PubMed]
6. Karakoy, M.; Gultepe, E.; Pandey, S.; Khashab, M.A.; Gracias, D.H. Silane surface modification for improved bioadhesion of esophageal stents. *Appl. Surf. Sci.* **2014**, *311*, 684–689. [CrossRef] [PubMed]
7. Lung, C.Y.K.; Sarfraz, Z.; Habib, A.; Khan, A.S.; Matinlinna, J.P. Effect of silanization of hydroxyapatite fillers on physical and mechanical properties of a bis-GMA based resin composite. *J. Mech. Behav. Biomed. Mater.* **2016**, *54*, 283–294. [CrossRef] [PubMed]
8. Lee, J.H.; Shin, H.; Rhee, K.Y. Surface functionalization of boron nitride platelets via a catalytic oxidation/silanization process and thermomechanical properties of boron nitride-epoxy composites. *Compos. Part B Eng.* **2019**, *157*, 276–282. [CrossRef]
9. Jiang, S.; Li, Q.; Zhao, Y.; Wang, J.; Kang, M. Effect of surface silanization of carbon fiber on mechanical properties of carbon fiber reinforced polyurethane composites. *Compos. Sci. Technol.* **2015**, *110*, 87–94. [CrossRef]
10. Yaghoubi, A.; Alavi Nikje, M.M. Silanization of multi-walled carbon nanotubes and the study of its effects on the properties of polyurethane rigid foam nanocomposites. *Compos. Part A Appl. Sci. Manuf.* **2018**, *109*, 338–344. [CrossRef]
11. Barczewski, M.; Matykiewicz, D.; Szostak, M. The effect of two-step surface treatment by hydrogen peroxide and silanization of flax/cotton fabrics on epoxy-based laminates thermomechanical properties and structure. *J. Mater. Res. Technol.* **2020**, *9*, 13813–13824. [CrossRef]
12. Goh, C.K.; Valavan, S.E.; Low, T.K.; Tang, L.H. Effects of different surface modification and contents on municipal solid waste incineration fly ash/epoxy composites. *Waste Manag.* **2016**, *58*, 309–315. [CrossRef] [PubMed]
13. En, S.; Nugay, N. Tuning of final performances of unsaturated polyester composites with inorganic microsphere/platelet hybrid reinforcers. *Eur. Polym. J.* **2021**, *37*, 2047–2053. [CrossRef]
14. Gohatre, O.K.; Biswal, M.; Mohanty, S.; Nayak, S.K. Effect of silane treated fly ash on physico-mechanical, morphological, and thermal properties of recycled poly(vinyl chloride) composites. *J. Appl. Polym. Sci.* **2021**, *138*, 50387. [CrossRef]
15. Xue, X.; Liu, Y.L.; Dai, J.G.; Poon, C.S.; Zhang, W.D.; Zhang, P. Inhibiting efflorescence formation on fly ash–based geopolymer via silane surface modification. *Cem. Concr. Compos.* **2018**, *94*, 43–52. [CrossRef]
16. Członka, S.; Sienkiewicz, N.; Strąkowska, A.; Strzelec, K. Keratin feathers as a filler for rigid polyurethane foams on the basis of soybean oil polyol. *Polym. Test.* **2018**, *72*, 32–45. [CrossRef]
17. Tan, S.; Abraham, T.; Ference, D.; MacOsko, C.W. Rigid polyurethane foams from a soybean oil-based Polyol. *Polymer* **2011**, *52*, 2840–2846. [CrossRef]
18. Bo, G.; Xu, X.; Tian, X.; Wu, J.; He, X.; Yan, Y. Synthesis and characterization of flame-retardant rigid polyurethane foams derived from gutter oil biodiesel. *Eur. Polym. J.* **2021**, *147*, 110329. [CrossRef]
19. Xi, W.; Qian, L.; Huang, Z.; Cao, Y.; Li, L. Continuous flame-retardant actions of two phosphate esters with expandable graphite in rigid polyurethane foams. *Polym. Degrad. Stab.* **2016**, *130*, 97–102. [CrossRef]
20. Qian, L.; Li, L.; Chen, Y.; Xu, B.; Qiu, Y. Quickly self-extinguishing flame retardant behavior of rigid polyurethane foams linked with phosphaphenanthrene groups. *Compos. Part B Eng.* **2019**, *175*, 107186. [CrossRef]
21. Gupta, A.; Badr, Y.; Negahban, A.; Qiu, R.G. Energy-efficient heating control for smart buildings with deep reinforcement learning. *J. Build. Eng.* **2020**, *34*, 101739. [CrossRef]
22. Członka, S.; Strąkowska, A.; Kairytė, A.; Kremensas, A. Nutmeg filler as a natural compound for the production of polyurethane composite foams with antibacterial and anti-aging properties. *Polym. Test.* **2020**, *86*, 106479. [CrossRef]
23. Członka, S.; Strąkowska, A.; Strzelec, K.; Kairytė, A.; Vaitkus, S. Composites of rigid polyurethane foams and silica powder filler enhanced with ionic liquid. *Polym. Test.* **2019**, *75*, 12–15. [CrossRef]
24. Bo, G.; Xu, X.; Tian, x.; Wu, J.; He, X.; Xu, L.; Yan, Y. Enhancing the flame retardancy for castor oil-based rigid polyurethane foams via silica aerogel. *J. Non. Cryst. Solids* **2021**, *562*, 120783. [CrossRef]
25. Acuña, P.; Lin, X.; Calvo, M.S.; Shao, Z.; Pérez, N.; Villafañe, F.; Rodríguez-Pérez, M.Á.; Wang, D.-Y. Synergistic effect of expandable graphite and phenylphosphonic-aniline salt on flame retardancy of rigid polyurethane foam. *Polym. Degrad. Stab.* **2020**, *179*, 109274. [CrossRef]
26. De Mello, D.; Pezzin, S.H.; Amico, S.C. The effect of post-consumer PET particles on the performance of flexible polyurethane foams. *Polym. Test.* **2009**, *28*, 702–708. [CrossRef]
27. Harikrishnan, G.; Singh, S.N.; Kiesel, E.; Macosko, C.W. Nanodispersions of carbon nanofiber for polyurethane foaming. *Polymer* **2010**, *51*, 3349–3353. [CrossRef]
28. Qi, Y.L.X.; Zhang, Y.; Chang, C.; Luo, X. Thermal, mechanical, and morphological properties of rigid crude glycerol-based polyurethane foams reinforced with nanoclay and microcrystalline cellulose. *Eur. J. Lipid Sci. Technol.* **2018**, *120*, 1700413. [CrossRef]
29. Sung, G.; Kim, J.H. Influence of filler surface characteristics on morphological, physical, acoustic properties of polyurethane composite foams filled with inorganic fillers. *Compos. Sci. Technol.* **2017**, *146*, 147–154. [CrossRef]
30. Zhou, X.; Sethi, J.; Geng, S.; Berglund, L.; Frisk, N.; Aitomäki, Y.; Sain, M.M.; Oksman, K. Dispersion and reinforcing effect of carrot nanofibers on biopolyurethane foams. *Mater. Des.* **2016**, *110*, 526–531. [CrossRef]

31. Husainie, S.M.; Deng, X.; Ghalia, M.A.; Robinson, J.; Naguib, H.E. Natural fillers as reinforcement for closed-molded polyurethane foam plaques: Mechanical, morphological, and thermal properties. *Mater. Today Commun.* **2020**, *27*, 102187. [CrossRef]
32. Kuźnia, M.; Magiera, A.; Jerzak, W.; Ziąbka, M.; Lach, R. Study on chemical composition of fly ash from fluidized-bed and convetional coal combustion. *Przemysł Chem.* **2017**, *96*, 1699–1703. [CrossRef]
33. Kuźnia, M.; Magiera, A.; Jerzak, W.; Ziąbka, M.; Zygmunt-Kowalska, B.; Kaczorek-Chrobak, K.; Pielichowska, K.; Szatkowski, P.; Benko, A.; Ziąbka, M.; et al. Fly Ash as an Eco-Friendly Filler for Rigid Polyurethane Foams Modification. *Materials* **2021**, *14*, 6604. [CrossRef] [PubMed]
34. Kuźnia, M. Reusing of fly ash from coal combustion in technology of polyurethane materials. *Przemysł Chem.* **2021**, *100*, 816–818. [CrossRef]
35. Sroka, J.; Rybak, A.; Sekula, R.; Sitarz, M. An Investigation into the Influence of Filler Silanization Conditions on Mechanical and Thermal Parameters of Epoxy Resin-Fly Ash Composites. *J. Polym. Environ.* **2016**, *24*, 298–308. [CrossRef]
36. Park, S.J.; Cho, S.H.; Seo, H.W.; Do Nam, J.; Suhr, J. Natural cork agglomerate enabled mechanically robust rigid polyurethane foams with outstanding viscoelastic damping properties. *Polymer* **2021**, *217*, 123437. [CrossRef]
37. Borowicz, M.; Paciorek-Sadowska, J.; Lubczak, J.; Czupryński, B. Biodegradable, flame-retardant, and bio-based rigid polyurethane/polyisocyanurate foams for thermal insulation application. *Polymers* **2019**, *11*, 1816. [CrossRef] [PubMed]
38. Członka, S.; Strąkowska, A.; Kairytė, A. Effect of walnut shells and silanized walnut shells on the mechanical and thermal properties of rigid polyurethane foams. *Polym. Test.* **2020**, *87*, 106534. [CrossRef]
39. Prociak, A.; Rokicki, G.; Ryszkowska, J. *Materiały Poliuretanowe*; PWN: Warsaw, Poland, 2016.
40. Badri, S.Z.K.H.; Ahmad, S.H. Production of a High-Functionality RBD Palm Kernel Oil-Based Polyester Polyol. *Appl. Polim. Sci.* **2001**, *81*, 384–389. [CrossRef]
41. Wolska, A.; Goździkiewicz, M.; Ryszkowska, J. Thermal and mechanical behaviour of flexible polyurethane foams modified with graphite and phosphorous fillers. *J. Mater. Sci.* **2012**, *47*, 5627–5634. [CrossRef]
42. Kairytė, A.; Kizinievič, O.; Kizinievič, V.; Kremensas, A. Synthesis of biomass-derived bottom waste ash based rigid biopolyurethane composite foams: Rheological behaviour, structure and performance characteristics. *Compos. Part A Appl. Sci. Manuf.* **2019**, *117*, 193–201. [CrossRef]
43. Mosiewicki, M.A.; Casado, U.; Marcovich, N.E.; Aranguren, M.I. Polyurethanes from tung oil: Polymer characterization and composites. *Polym. Eng. Sci.* **2009**, *49*, 685–692. [CrossRef]
44. Barczewski, M.; Kurańska, M.; Sałasińska, K.; Michałowski, K.; Prociak, A.; Uram, K.; Lewandowski, K. Rigid polyurethane foams modified with thermoset polyester-glass fiber composite waste. *Polym. Test.* **2020**, *81*, 106190. [CrossRef]
45. Jiao, L.; Xiao, H.; Wang, Q.; Sun, J. Thermal degradation characteristics of rigid polyurethane foam and the volatile products analysis with TG-FTIR-MS. *Polym. Degrad. Stab.* **2013**, *98*, 2687–2696. [CrossRef]
46. Kuźnia, M.; Magiera, A.; Pielichowska, K.; Ziąbka, M.; Benko, A.; Szatkowski, P.; Jerzak, W. Fluidized bed combustion fly ash as filler in composite polyurethane materials. *Waste Manag.* **2019**, *92*, 115–123. [CrossRef] [PubMed]
47. Duan, B.; Wang, Q.; Wang, X.; Li, Y.; Zhang, M.; Diao, S. Flame retardance of leather with flame retardant added in retanning process. *Results Phys.* **2019**, *15*, 102717. [CrossRef]
48. Cao, Z.J.; Dong, X.; Fu, T.; Deng, S.B.; Liao, W.; Wang, Y.Z. Coated vs. naked red phosphorus: A comparative study on their fire retardancy and smoke suppression for rigid polyurethane foams. *Polym. Degrad. Stab.* **2017**, *136*, 103–111. [CrossRef]

MDPI AG
Grosspeteranlage 5
4052 Basel
Switzerland
Tel.: +41 61 683 77 34

Energies Editorial Office
E-mail: energies@mdpi.com
www.mdpi.com/journal/energies

Disclaimer/Publisher's Note: The title and front matter of this reprint are at the discretion of the Guest Editors. The publisher is not responsible for their content or any associated concerns. The statements, opinions and data contained in all individual articles are solely those of the individual Editors and contributors and not of MDPI. MDPI disclaims responsibility for any injury to people or property resulting from any ideas, methods, instructions or products referred to in the content.

www.ingramcontent.com/pod-product-compliance
Lightning Source LLC
LaVergne TN
LVHW072353090526
838202LV00019B/2537